Routledge R

T0228050

Religion in Evolution

First published in 1906, these four lectures were originally delivered in the Vacation Term for Biblical Study at Cambridge. Evidence is derived from the native tribes of Australia in particular, whom Jevons compares for his discussion. The first lecture considers whether religion has evolved from, or was preceded by a pre-religion, or non-religious, stage of humankind. The following lectures discuss the science of religion and the theory of Evolution, and the relationship between the evolution of religion and the philosophy of religion. This is a fascinating text that will be of particular value to students interested in the history and development of religion in general.

Religion in Evolution

F. B. Jevons

Routledge
Taylor & Francis Group

First published in 1906
by Methuen & Co.

This edition first published in 2015 by Routledge
2 Park Square, Milton Park, Abingdon, Oxon, OX14 4RN
and by Routledge
711 Third Avenue, New York, NY 10017

Routledge is an imprint of the Taylor & Francis Group, an informa business

© 1906 Methuen & Co.

Publisher's Note
The publisher has gone to great lengths to ensure the quality of this reprint but points out that some imperfections in the original copies may be apparent.

Disclaimer
The publisher has made every effort to trace copyright holders and welcomes correspondence from those they have been unable to contact.

A Library of Congress record exists under LC control number: 07037990

ISBN 13: 978-1-138-81490-5 (hbk)
ISBN 13: 978-1-315-74711-8 (ebk)
ISBN 13: 978-1-138-81493-6 (pbk)

RELIGION IN EVOLUTION

BY

F. B. JEVONS, Litt.D.

METHUEN & CO.
36 ESSEX STREET W.C.
LONDON

First Published in 1906

CONTENTS

The first appearance, in Time, of consciousness upon earth —or of Religion—is a question the answer to which must be furnished by the theory of Evolution. Whether there is a principle of consciousness underlying the process of Evolution generally, and the Evolution of Religion in particular— a principle which even now, as a principle of Religion, manifests itself but imperfectly—is a question for Philosophy. The question dealt with in this Lecture is whether Religion has been evolved out of, or preceded by, a non-religious or pre-religious stage in the history of man. Such a stage has been supposed to have been discovered amongst the Australian black-fellows, some of whom believe in an All-father, "the father of all of us," "our father." This belief, it is said, does not, but might easily have amounted to Monotheism. The question therefore is whether this belief is a decline from, or a stage on the way to, Monotheism. It is held by the S.E. tribes of Central Australia, who are socially more advanced than those of the N. But there are indications that it was held once by the latter; and, if so, then the N. tribes are further away from the original beliefs of the common ancestors of the S.E. and the N. tribes than the S.E. tribes are. This inference, that the S.E. and the N. tribes have both declined, the latter more, the former less, from an earlier Monotheism, is confirmed by the parallel afforded by the negroes of W. Africa. The similarity between Africa and Australia in this respect suggests that we have here to do with a general tendency. If so, then a pre-religious stage in the history of man cannot yet be said to have been satisfactorily proved.

v

CONTENTS

CONTENTS

vii

CONTENTS

tion which experience has no constraining power to compel us to answer either way: some people have answered it first one way and then the other. We are in fact free to take either answer: it is a matter of Will. And to maintain the position once taken up requires a constant exercise of Will-power—for you are still free. That exercise of the power or the Will to believe is Faith; and Faith is not purely intellectual but emotional, and the emotion is Love. Either we do or we do not feel God's love for us, and our own gratitude for it. That is a question each must answer for himself; and the answer leaves no doubt whether the existence of God is a fact of experience or a mere assumption.

PREFACE

THESE four lectures were delivered in the Vacation Term for Biblical Study at Cambridge, and are printed at the request of those who heard them.

In Lecture I. I accept the statement of Mr Howitt in his "Native Tribes of South-east Australia" that the South-eastern tribes who believe in an All-father are socially more advanced than the Northern tribes, who, according to Messrs Spencer and Gillen, have no "belief of any kind in a supreme Being who rewards or punishes the individual according to his moral behaviour." At the time of writing I had not seen Mr A. Lang's letter to *Folk-Lore* (xvi. 2, pp. 221-224), in which he argues, against Mr Howitt, that the majority of the South-eastern tribes "are in the more primitive form of social organisation." I am not concerned to take sides on

PREFACE

this question, as the question, whichever way it is settled, does not affect my argument, the basis of which is that social or political progress does not necessarily imply or entail religious development, or even prevent religious decay; in fact, social development and religious development may vary directly or inversely, and the direction of the movement of either can only be ascertained by observation, not by inference from the direction in which the other moves. The important point is that the Northern tribes, in Mr Lang's opinion, "have almost sloughed off the belief" in the All-father, not that they never had it; and to that opinion I subscribe.

Whether there ever was a pre-religious stage in the development of man is an open question. Mr Frazer, in the extract from the forthcoming third edition of the "Golden Bough," which he gives in the *Fortnightly Review* (No. cccclxviii. N.S., pp. 162-172), does not make his opinion on this question,

PREFACE

so far as the aborigines of Australia are concerned, quite clear. He begins by saying that Religion, in his sense of the word, seems to be "nearly unknown" amongst them; he ends by saying that "if the Australian aborigines had been left to themselves they might have evolved a native Religion." The implication of these last words seems rather to be that amongst the Australian aborigines Religion is not "nearly unknown" but actually unknown—that there is or has been no native religion. It is, of course, a perfectly competent position to take up that, in the existing state of our knowledge, we are not justified in treating the point as decided: and that may be the real nature of Mr Frazer's apparent indecision on the point. On the other hand, if we are to press the words of the passage at the end of his article, and to understand them to mean that there was no native religion in Australia, then Mr Frazer's theory "that in the history of mankind Religion has been preceded by magic"

PREFACE

is confirmed—if there was indeed no native Religion in Australia. But it is of great interest to all students of the Science of Religion to know what position on this point Mr Frazer takes up; and his article in the *Fortnightly Review* leaves it uncertain whether he does or does not regard it as settled that there was no native Religion in Australia, and as therefore proved that in this case "Religion has been preceded by magic."

RELIGION IN EVOLUTION

I

THE theory of the descent of man from a non-human ancestor is generally accepted by those who are qualified to judge the evidence on which it is based. And by those who accept it the evolution of Religion from antecedent phenomena which were non-religious will seem *a priori* probable, even if the evidence at present at our disposal does not seem conclusive on the point. There are indeed difficulties of a philosophical kind, analogous to the difficulty of understanding how consciousness can be supposed to have been evolved in any sense out of unconscious matter—how matter which is known only as the object of thought, as the object of which a thinking subject is aware, can exist or have existed save as the object of thought, as the object of which a conscious mind or spirit is aware.

A   I

RELIGION IN EVOLUTION

And those who are alive to these difficulties will probably feel that they stand seriously in the way of any attempt to exhibit Religion as evolved from antecedent phenomena of a non-religious kind. Feeling these difficulties to be serious, some of us may incline to draw a distinction between the first appearances in which an underlying principle manifests itself and the principle itself. Thus the principle underlying the appearance of evolution may be a principle of thought or consciousness, or moral consciousness, which even as yet has but very imperfectly manifested itself, and before its first appearance, of course, had not manifested itself at all. But though then it had not manifested itself, and though now it manifests itself but imperfectly, still it was and is the underlying principle of evolution, revealing itself in evolution. If there were antecedent phenomena, if there were phenomena which, apprehended under the form of Time, preceded the first appearance of intellectual consciousness, or religious consciousness, then those phenomena, out of which Religion, on the theory of evolution, was evolved, do not, and *ex hypothesi* did not,

2

constitute Religion; nor is Religion resolved into them, if we should succeed in going back to its first appearance and in re-constituting the state of things in which it made its first appearance.

Now, to the philosophical question I may have occasion to revert hereafter. It is obviously different from the question of fact, whether as a matter of fact Religion has been evolved out of or preceded by a non-religious or pre-religious stage. That is a question of the evolution of Religion; and it is with the evolution, not with the philosophy, of Religion that I shall be concerned in this lecture. Indeed it is precisely with this question of fact, viz. whether Religion has been evolved out of, or has been preceded by, a non-religious or pre-religious stage, that I shall choose to deal. Or, to be yet more precise, it is with one particular answer to this question of fact that I shall deal in this chapter. The particular answer is that given by Mr A. W. Howitt in his recently published work, entitled "The Native Tribes of South-East Australia." I need hardly say that there is no man living who has such an acquaintance with those tribes as Mr Howitt, and no man who can speak of their modes of thought and ways of life with greater authority than he. Now, the outcome of Mr Howitt's forty years' acquaintance with these tribes and work amongst them is the conclusion that—in his own words—"it cannot be alleged that these aborigines have consciously any form of Religion" (p. 507). If this conclusion of Mr Howitt's be correct, then we actually have at the present day in the British Empire, tribes not merely in a non-religious stage, but in a pre-religious stage. "Their beliefs," he says, "are such that, under favourable conditions, they might have developed into an actual Religion." The kind of Religion into which their beliefs might have developed is, according to Mr Howitt, monotheism. We have therefore in the beliefs of these tribes, if Mr Howitt is right, the antecedent phenomena out of which Religion might have been—though in Mr Howitt's view it was not—evolved by these tribes—phenomena which, in his view, do not constitute Religion, though they might well have been followed by the first appearance of Religion. And that Religion, in his view, would have been a monotheism. But in saying this, he is most anxious to have it understood that he has not been swayed by any considerations of a theological or non-scientific character : " In saying this I must guard myself from being thought to imply any primitive revelation of a monotheistic character. What I see is merely the action of elementary thought reaching conclusions such as all savages are capable of, and which may have been at the root of monotheistic beliefs " (*ib.*).

What, then, is the evidence which indicates that these South-Eastern tribes, though they have no conscious form of Religion, were on the direct line for developing monotheism, rather than polytheism, or ancestor-worship, or animal-worship ? To begin with, we must notice that amongst these tribes there is, in Mr Howitt's words, " a universal belief in the existence of the human spirit after death " (p. 440). Very naturally, the human spirits which continue to exist after death are supposed to exist in much the same way as before death : they live in the sky-country in the same tribal organisation as on this earth ; and as they have a Head Man here, so they have a Head Man there. Now, it would seem that, though Mr Howitt speaks of the head of the tribe of the dead as a Head Man, the natives themselves do not describe him by the same word. The being in the sky-country doubtless exercises many of the functions and has many of the attributes enjoyed by the person who amongst the natives occupies the official position of Head Man; and Mr Howitt's view of this being in the sky-country is that he is supposed to be what the Head Man of a tribe is in this world. Mr Howitt infers this from his wide knowledge of the natives and their beliefs. " Combining," he says, "the statements of the legends and the teachings of the ceremonies, I see, as the embodied idea, a venerable, kindly Head Man of a tribe, full of knowledge and tribal wisdom, and all-powerful in magic, of which he is the source, with virtues, failings, and passions, such as the aborigines regard them " (p. 500). This being in the sky-country is, Mr Howitt tells us, known generally amongst these tribes as "our father," or "father of all of us." Now, it is true that the official Head Man is not officially addressed or spoken of as "our father," or "father of all of us"; but Mr

6

RELIGION IN EVOLUTION

Howitt says "it is not a long stretch to the idea of the All-father of the tribe, since it is not uncommon, indeed I may go so far as to say that it is, in my experience, common, to address the elder men as father" (p. 507). Now, it would be very natural for us to imagine that this "father of us all" is regarded by the natives as a divine being; but Mr Howitt is satisfied that we should be wrong in so doing. "It is most difficult," he says, "for one of us to divest himself of the tendency to endow such a supernatural being with a nature *quasi*-divine, if not altogether so—divine nature and character" (p. 501). Indeed, as a matter of fact, various explorers and travellers in Australia, whom Mr Howitt quotes, have inferred, from the fact that the natives believe in this "supernatural being," the conclusion that these blacks believe in "a supreme being or deity." But the conclusion is felt by Mr Howitt to be undoubtedly wrong. He says, "in this being, although supernatural, there is no trace of a divine nature. All that can be said of him is that he is imagined as the ideal of those qualities which are, according to their standard, virtues worthy of being imitated. Such would be a man who is skilful in the use of weapons of offence and defence, all-powerful in magic, but generous and liberal to his people, who does no injury or violence to anyone, yet treats with severity any breaches of custom or morality. Such is, according to my knowledge of the Australian tribes, their ideal of Head Man, and naturally it is that of the *Biamban*, the master in the sky-country. Such a being from *Bunjil* to *Baiame*, is *Mami-ngata*, that is, 'our father'; in other words, the 'All-father of the tribes'" (p. 507). Finally, there is one more important fact to be noticed in support of Mr Howitt's view. If this All-father were really felt by the blacks to be a supreme being or deity, we should expect him to be worshipped. "But," says Mr Howitt, "there is not any worship of *Daramulun*" (p. 507). It is indeed the case that a figure of clay, an image of Daramulun, is made, and that there are dances round it. These facts, however, are regarded by Mr Howitt as showing not that the "aborigines have consciously any form of Religion," but that "under favourable conditions they might have developed into an actual Religion." "There is not any worship of *Daramulun*," he says, "but the dances round the figure of clay and the invocating of his name by the medicine-men certainly might have led up to it" (p. 508).

Now, it may appear to some of us that the tribes to which Mr Howitt refers are not merely on the verge of passing from the pre-religious to the religious stage, but have actually passed it. And to a certain extent, if we take up that position, we may fortify ourselves with quotations from Mr Howitt. Thus by one tribe this supernatural being "is said to have made all things on the earth and to have given to men the weapons of war and hunting, and to have instituted all the rites and ceremonies which are practised by the aborigines, whether connected with life or death" (p. 488). Another tribe speaks of him "with the greatest reverence. He was said to have made the whole country, with the rivers, trees, and animals. He gave to the blacks their laws" (p. 489). According to yet another set of tribes, he "was the maker of the earth, trees, and men" (p. 492). According to the belief of another tribe, he is "the

maker who created and preserves all things"
(p. 494). Other aborigines say he told them
what to do, "and he gave them the laws
which the old people have handed down from
father to son to this time" (p. 495). Elsewhere it is believed that "*Tharamulun* can
see people and is very angry when they do
things that they ought not to do, as when
they eat forbidden food" (p. 495).

I think that if we pressed these passages
that I have quoted we might maintain with a
certain degree of plausibility, at the least, that
the tribes in question are not merely on the
verge of passing from the pre-religious to the
religious stage, but have actually passed it.
And though the absence of worship, which I
have already mentioned, and the absence, still
more, of prayer, may make us hesitate to go
further than Mr Howitt allows us, still in
principle, whether these tribes are on the
verge or have passed it makes little difference.
If they did not take the step, at any rate on
this theory of the origin of Religion, other
peoples in other parts of the world did take
it ; and so we have before our eyes, as it were,
the actual process in actual working whereby

RELIGION IN EVOLUTION

Religion is evolved from or supervenes upon antecedent phenomena of a non-religious kind. That is in effect the theory. Here we have certain Australian tribes on the line between Religion and non-religion ; and the view is submitted that they have advanced to this point from the region of non-religion. Now, there must be some reason for assuming that they have progressed to this point from the region of non-religion rather than that they have declined to it from some more conscious form of Religion; and that reason is given by Mr Howitt. He says, "that part of Australia which I have indicated as the habitat of tribes having this belief [*i.e.* the belief in "our father"] is also the area where there has been the advance from group marriage to individual marriage, from descent in the female line to that in the male line ; where the primitive organisation under the class system has been more or less replaced by an organisation based on locality ; in fact, where those advances have been made to which I have more than once called attention in this work " (p. 500). There, then, is the reason : these tribes have advanced in social organisation, therefore probably their movement in matters affecting Religion has also been one of progress and advance. Now, it is at least conceivable, and I do not at this moment put it forward as more than a conceivable view, that the general movement in matters affecting Religion has been one of retrogression, both in these tribes whose social organisation is more evolved and in those other tribes whose social organisation has been less evolved. On this view, it would be natural enough that tribes which actually progressed socially would resist religious deterioration more successfully than tribes which were incapable even of social advance. Indeed some of us might go so far as to suggest that it was precisely because the one set of tribes clung more faithfully than the other to their religious traditions that they made social progress; and that if the other tribes made no social progress, it was just because they had declined from the religious point of view.

But have they declined from the religious point of view? Whether they have or have not been the victims of a retrogression in Religion, it is at any rate clear from Mr Howitt's words that the tribes which relatively have made no social progress are in a different position as regards religious beliefs to the tribes in which social advance has been made. In the socially progressive tribes, he says, " a belief exists in an anthropomorphic supernatural being [the All-father] who lives in the sky, and who is supposed to have some kind of influence on the morals of the natives. No such belief seems to obtain in the remainder of Australia, although there are indications of a belief in anthropomorphic beings inhabiting the sky-land " (p. 500).

I propose now, therefore, in order to gain some information about the beliefs which obtain in the remainder of Australia, to turn to a work of the very highest authority : I mean Messrs Spencer and Gillen's " Northern Tribes of Central Australia." These Northern tribes are, Messrs Spencer and Gillen say, "savages who have no idea of permanent abodes, no clothing, no knowledge of any implements save those fashioned out of wood, bone, and stone ; no idea whatever of the cultivation of crops, or of the laying in of a supply of food to tide over hard times ; no word for any number beyond three, and

no belief in anything like a supreme being"
(p. xiv). "We know," they say, "of no tribe
in which there is a belief of any kind in a
supreme being who rewards or punishes the
individual according to his moral behaviour,
using the word moral in the native sense"
(p. 491). Thus these Northern tribes are
very different from Mr Howitt's South-Eastern
tribes, who believe that laws were given to
them by "our father," and that "he is very
angry when they do things that they ought
not to do." Now, I think that anyone who
knew nothing more of the subject than the
quotations I have given in this paper, and
who was inclined to believe that the religious
ideas, like the social organisation, of Mr
Howitt's South-Eastern tribes, were evolved
from, and an advance upon, those of the
Northern tribes, would be led to expect that
the Northern tribes had not yet attained to
the conception of an anthropomorphic supernatural All-father living in the sky; or that,
supposing they had, at any rate he was not
imagined to have anything to do with the
morals of the natives. Yet this expectation
would not be altogether correct. Both the

RELIGION IN EVOLUTION

Northern and the South-Eastern tribes have initiation-ceremonies or mysteries, from which the women are jealously excluded. At these ceremonies the simple moral rules or laws of the natives are solemnly impressed upon the boys who are initiated. Messrs Spencer and Gillen say of their tribes, "So far as the inculcation of anything like moral ideas is concerned, this, such as it is, may be said to take place always in connection with initiation" (p. 502). Now, the women and children are taught to believe that, on the occasion of these mysteries, "a spirit takes the boy out into the bush, enters the body of the boy, and brings him back again initiated" (pp. 497, 499). If, therefore, these were all the facts we had to go on, we should be in this position : we should know that amongst Mr Howitt's South-Eastern tribes a boy, when initiated, was taught to believe in an anthropomorphic, supernatural being who lived in the sky, was the creator and preserver of all things, was the giver of moral laws, who was very angry with people if they did what they ought not to do, and who finally was "the father of all of us."

RELIGION IN EVOLUTION

We should further know that amongst the
Northern tribes the inculcation of moral ideas
took place at the initiation-ceremonies; and
that the women and children believed that
an anthropomorphic, supernatural being played
a part in them. I think, then, that we should
go on to infer that the women and children
of the Northern tribes were not far wrong,
and that the boy was taught in the Northern
tribes what a boy in the South-Eastern tribes
was taught, viz. to believe in "the father
of all of us." But there we should be wrong:
what happens at the initiation as a matter of
fact is (in Messrs Spencer and Gillen's words),
that "he then learns that the spirit creature
whom, up to that time, as a boy, he has
regarded as all-powerful, is merely a myth,
and that such a being does not really exist,
and is only an invention of the men to
frighten the women and children" (p. 492).
From these words it is clear that as a boy
he was taught that at the mysteries he would
be initiated by an all-powerful spirit creature;
that the men spread abroad the story, or
allowed it to be spread, that the spirit
appeared and performed the initiation (which

was supposed to consist, in Messrs Spencer and Gillen's words, "in cutting out all his insides and providing him with a new set," p. 498); and that the boy, when he learnt at the initiation that he had been defrauded, became interested in keeping up the fraud.

The case, then, as we have it now, is that at the initiation ceremonies the men of the South-Eastern tribes believe, and teach their boys the belief, in the All-father, the giver of such moral laws as the black fellows have; whereas the men of the Northern tribes teach their boys "that such a being does not exist and is only an invention of the men to frighten the women and children." The question then inevitably rises, though I have not yet seen it stated or discussed, which of these two doctrines is the earlier. For my own part, I see no possibility of doubt. If the belief in the All-father is supposed to be the original, or the earlier, belief, it might easily degenerate into a mere survival, when faith in it, for whatever reason, was lost. Naturally the men who were initiated into the mysteries would not, in the later stage of their development any more than in the earlier, give them

B 17

away : indeed the secret would all the more jealously be kept. On the other hand, if we were to hold that disbelief in the supposed supernatural being was the earlier stage, it would be difficult to imagine how belief grew out of it ; and, as a matter of fact, disbelief pre-supposes the existence of the belief—the belief is there, held by some persons and rejected by others ; it could not be disbelieved before it existed. It must have existed first and then have come to be disbelieved. That we might safely say, if we had only Mr Howitt's account of the South-Eastern tribes to go upon. But fortunately we are not in the position of having to say as a matter of inference and conjecture that the belief which is found amongst the South-Eastern tribes must have existed amongst the Northern tribes before the Northern tribes could come to disbelieve it : we have Messrs Spencer and Gillen's evidence for it that it does exist amongst the Kaitish tribe. They say "amongst the Kaitish we meet with a spirit individual named Atnatu, the beliefs with regard to whom are different from those concerning Twanyirika, and are peculiar to

this tribe. This Atnatu . . . made himself
and gave himself his name." He lives up
beyond the sky and "he let down everything which the black fellow has — spears,
boomerangs, tomahawks, clubs, everything in
fact," but the women "know nothing about
Atnatu" (pp. 498, 499).

When Messrs Spencer and Gillen wrote
and published their book, Mr Howitt's work
had not appeared : the Kaitish beliefs were
without parallel amongst the Northern tribes,
and it was not unreasonable to regard their
isolated set of beliefs as something sporadic
and peculiar. Amongst all the other Northern
tribes the spirits spoken of were, as Messrs
Spencer and Gillen say, "merely bogies to
frighten the women and children and keep
them in a proper state of subjection" (pp.
502, 503). It was not unnatural, therefore,
for Messrs Spencer and Gillen, having only
before them the evidence afforded by the
Northern tribes, to say there does not "appear to be any evidence which would justify
the hypothesis that the present ideas with
regard to them [*i.e.* these spirits], are the
result of degradation" (p. 508). But since

the appearance of Mr Howitt's work the evidence that the ideas of the Northern tribes are the result of degradation, and are a degradation from the South-Eastern tribes' belief in the All-father, has been decisive on the point.

This supplementary evidence, so valuable and conclusive, is a good example of the value of the comparative method in the study of Religion. A fact which, taken by itself, is puzzling and incomprehensible, becomes intelligible and the key to the situation when the method of comparison is set to work, and shows the fact to exist elsewhere in what is evidently its right relation to the circumstances. I propose, therefore, now to employ the comparative method again, and I hope by doing so to show that the facts which we have been considering are not merely of interest to those who happen from some inscrutable reason to concern themselves with the beliefs of these Australian black fellows, almost, if not quite, the lowest of the human race. Nor are these facts peculiar to the Australian tribes: they recur amongst a people with whom they can have hardly come in contact, and I will ask

RELIGION IN EVOLUTION

you to turn with me to a book written by a missionary who has had more than forty years' experience of the natives of whom he writes. It is " Fetichism in West Africa," written by the Rev. Robert Nassau.

"Among the negro tribes of the Bight of Benin and the Bantu of the region of . . . what is now the Kongo-Français, there was a power," says Mr Nassau, "known variously as Egbo, Ukuku, and Yasi, which tribes, native chiefs, and headmen of villages invoked as a court of last appeal, for the passage of needed laws, or the adjudication of some quarrel which an ordinary family or village council was unable to settle. . . . Egbo, Ukuku, Yasi was a secret society composed only of men ; boys being initiated into it about the age of puberty. Members were bound by a terrible oath, and under pain of death, to obey any law or command issued by the spirit under which the society professed to be organised " (pp. 139, 140) ; "recalcitrants would submit instantly and in terror of Ukuku's voice. They (the men) taught their little children, both girls and boys, that the voice belonged to a spirit which ate people who disobeyed him.

RELIGION IN EVOLUTION

When the society walked in procession they were preceded by runners who warned all on the path of the coming of the spirit. Women and children hastened to get out of the way; or, if unable to hide in time, they averted their faces. The penalty when a woman even saw the procession was a severe beating" (p. 141). Mr Nassau speaks from personal acquaintance of the Egbo, Ukuku, and Yasi of the Negro tribes and of the Bantu in the Kongo-Français. But these secret societies are found over a much wider area. He says "there is also in the Gabun region of the equator, among the Shekani, Mwetyi; among the Bakele, Bweti; among the Mpongwe-speaking tribes, Indo and Njĕmbĕ; and Ukuku and Malinda in the Batanga regions" (p. 248). Now, of these secret societies, or mysteries, or organisations, he says: "All these societies had for their primary object the good one of government" (p. 248); and elsewhere, "like all government intended for the benefit and protection of the governed, Ukuku, when it happened to throw its power on the side of right, was occasionally an apparent blessing" (p. 145). He quotes from a Sierra Leone newspaper the statement that "these institutions are connected with, and govern the agencies that work in, the sociology of all communities, such as the marriage laws, the relation of children to parents and of sex to sex, social laws, the position of eldership and the deference to be paid to age and worth, native herbs and medicines" (p. 146). How closely, then, the functions of these African organisations resemble those of the Australian organisations with which I am comparing them will be seen if I quote from Messrs Spencer and Gillen the injunctions which are imposed upon the Australian novice at the time of his initiation. They are, "speaking generally, the following :—(1) That he must obey his elders and not quarrel with them; (2) that he must not eat certain foods " [this restriction, though not mentioned in the Sierra Leone newspaper, is as widely spread in Africa as in Australia]; "(3) that he must not attempt to interfere with women who have been allotted to other men, or belong to groups with the individuals of which it is not lawful for him to have marital relations; (4) that he must on no account reveal any of the secret matters of the totems to the women and children " (p. 503).

The next point that I wish to make in this parallel between the African and the Australian mysteries is that as in Australia the belief in the All-father or "father of all of us" is preserved amongst some tribes, but in other tribes survives only as a device of the men to frighten the women and children, or has died out altogether ; so too in Africa, in some cases, especially, as Mr Nassau says, "among the tribes of the interior, where foreign government is as yet only nominal" (p. 248), the belief in the spirit is genuine and operative, while in others the natives who carry on the organisation know, in Mr Nassau's words, that "the whole proceeding was an immense fiction" (p. 140), and that "they helped to carry on a gigantic lie" (p. 141). Both stages in the history of the institution are portrayed by Mr Nassau. I wish I had space to quote in full the account which he took down from the lips of a native who told him "freely what happened when he was initiated as a lad"; how "early one morning the voices of the elders were heard in the street, ' Malanda has

24

come!' The women and girls were frightened. They knew they were not to look at Malanda. And we lads were oppressed with a vague dread that subdued us from our usual boisterous plays. We knew the name 'Malanda.' It was a power; it was mysterious. Mystery is a burden : it might be for good or for evil." Some twenty lads were made to sit upon a log facing the sun. "We were told to throw our heads back, bending our necks to the point of pain, and to stare with unblinking eyes at the sun. As the sun mounted all that morning, hot and glaring, toward the zenith, we were sedulously watched to see that we kept our heads back, arms down, and eyes following the burning sun in its ascent. My throat was parched with thirst. My brain began to whirl, the pain in my eyes became intolerable, and I ceased to hear; all around me became black, and I fell off the log. As each one of us thus became exhausted, we were blindfolded and taken to that house. On reaching it, still blindfolded, I knew nothing that was there. I smelled only a horrible odour. It was useless to resist, as they began to beat me with rods. My outcries only brought severer blows. I perceived that submission lightened their strokes. When finally I ceased struggling or crying, the bandage was removed. The horror of that headless corpse standing extending its rotting arms towards me, and the staring glass eyes of the image overcame me, and I attempted to flee. That was futile. I was seized and beaten more severely than before, until I had no will or wish, but utter submission to the will of whatever power it might be into whose hands I had fallen" (p. 323). But I must not pursue the quotation any further, or describe his twenty days' experiences in that prison : he was "entrusted with a secret to which younger lads were not admitted, and from which all of womankind were debarred" (p. 324) ; "although still confined, I did not feel that I was a prisoner; I was deeply interested in seeing and taking part in this great mystery" (p. 325). Of the native who told him this story, Mr Nassau says he was "without even a pretence of Christianity ; at heart a heathen, though a member of the Roman Catholic Church, into which he consented to be baptised as the means of obtaining in marriage his wife, who had been raised in that Church" (p. 320). On the other hand, to present an account of these African mysteries when they have degenerated into conscious fraud, I must quote the case of a convert made by Mr Nassau's Mission. " It had occurred in the early history of the Mission that one young man, Ibia, a freeman, member of a prominent family, had felt that in breaking away from heathenism and becoming a Christian he should cast off the very semblance of any connection with evil, or even tacit endorsement of it. He knew the society was based on a great falsehood. As a lad he had believed Ukuku was a spirit ; on his initiation he had found that this was not so ; but, loyal to his heathenism and his oath, he had assented to the lie, and had assisted in propagating it. He was known for the fearlessness of his convictions ; and in his conversion he to a rare degree emerged from all superstitious beliefs. Few emerge so utterly as he. He therefore publicly began to reveal the ceremonies practised in the Ukuku meetings. At once his life was in danger." Many attempts were made upon it. But "he came through his fiery trial strong, and his life has since become that of a reformer. He became the Rev. Ibia j'Ikĕngĕ, member of Corisco presbytery and pastor of the Corisco church; and Ukuku has long since ceased to exist as a power on the island" (p. 145). When Mr Nassau has occasion to mention this and similar instances to the men, they wince and say, "Don't speak so loud; the women will hear you." Thus in Africa, as in Australia, the original belief in the All-father has in some cases been lost; the ceremonies in which it originally found expression survive; and then the belief which originally was genuine becomes, as Messrs Spencer and Gillen say, a mere bogey "to frighten the women and children and keep them in a proper state of subjection."

In Africa and Australia alike these mysteries —even those of them which enshrine a genuine religious belief—when they are in the charge of men, eventually become known. But as in Athens there were mysteries, the Thesmophoria, to which women alone were admitted, so in Africa there are mysteries, Njĕmbĕ, to which women alone are admitted. And these mysteries are mysteries still : the women keep the secret. " Nothing is known of their rites," Mr Nassau says ; " the entire process so beats down the will of the novices and terrorises them, that even those who have been forced into it against their will, when they emerge at the close of the rite, most inviolably preserve its secrets, and express themselves as pleased " (p. 250). " It is remarkable," he says, " how well the secrets of the society are kept. No one has ever been induced to reveal them. Those who have left the society and have become Christians do not tell. Foreigners have again and again tried to bribe, but in vain. Traders and others have tried to induce their native wives to reveal ; but these women, obedient to any extent on all other matters, maintain a stubborn silence " (p. 254). Of these real mysteries, therefore, I can say nothing more, except that on the last day of them the women go round begging " gifts of rum, tobacco, plates, and cloth. In a civilised religious worship," Mr Nassau says, " this would be the taking up of the collection " (p. 254). The practice, I may say, is not known to me in connection with any mysteries confined to men, and is probably a spontaneous manifestation of woman's natural capacity for business.

If now we look to see to what point the argument has brought us thus far, it is this : in places so far distant from one another and so unlike each other as Australia and Western or Equatorial Africa, we find that boys are initiated into certain rites which are mysteries in so far as women, children, and strangers are excluded, and that on initiation they are taught certain beliefs respecting Religion and morality. We further find these mysteries in each country in two stages : in one, which I suggest was the earlier, we find existing a belief in a spirit, who made and preserves all things, and who gave the natives the moral laws which they recognise and on which their social organisation is based. In the other stage of the history of these mysteries we find this spirit regarded as merely a bogey to frighten the women and children, and having nothing to do with rewarding or punishing the individual according to his moral behaviour—using the word moral in the native sense. But whereas in Australia the men of the Northern tribes have all, according to Messrs Spencer and

RELIGION IN EVOLUTION

Gillen, come to understand that this All-father is a myth and merely a bogey, in Western Africa, even where the mysteries have dwindled or disappeared, the belief in this spirit, as God, has not disappeared. According to a paper read by M. Allégret at the International Congress on the History of Religions, in 1904, the Fan people, one of the most important groups of the great Bantu family, still believe that there exists "a superior being, Nzame, creator of all things, who lives far away, and who is still capable of exercising his power on occasion. He it is who, in a sense, governs the world; but the conception of his presence and activity is rapidly perishing. 'We all know,' they say, 'that God exists: He it is who made us.'" And not only the Fan, but, says M. Allégret, "all the Bantu peoples with whom I have had to do in this part of Western Africa designate the Supreme Being by the same name" (*Revue de l'Histoire des Religions*, l. 2, 223-5). But, says M. Allégret, "these religious ideas have practically no influence now on the ordinary life of the Fan" (p. 226). And that is a very important statement. Mr Andrew Lang, in

31

RELIGION IN EVOLUTION

his work entitled " The Making of Religion," has collected evidence showing that the belief in a superior or supreme being is found widely spread among the lower races of the whole world. But he points out that, as a rule, no cult or ritual goes with the belief. So that what M. Allégret says of the Fan belief seems to be generally true of this belief as it now exists among the lower races : "these religious ideas have practically no influence now on ordinary life."

The question, however, naturally arises, whether these religious ideas were always, as they now are mostly, without influence on ordinary life ; or whether they had it originally and have since lost it. It is a question on which we are in the dark as regards most of the races in whom the belief survives, for they have no history. But when we turn to those tribes about whom we know a little more, we find that these beliefs are not, or have not been, without influence on the ordinary life of those who hold them. Both in Australia and in Western Africa, as we have seen, the ordinary morality of the native is under the sanction of the being in whom the boys at the mysteries are taught to believe — as long as they are taught to believe. It is not, therefore, wholly unreasonable to surmise that the belief in a superior or supreme being elsewhere was originally bound up with, and gave its sanction to, the morality of the native : Religion and morality are thus closely united in the case of tribes in Australia and Western Africa which stand at the bottom of the scale of social evolution, and the fact weighs in favour of those who hold that the connection is original. Its weight will naturally be regarded as considerable by those who feel the ultimate basis for morality to be the desire to do the will of God.

It so happens that in Australia among some tribes we find Religion and morality divorced, and amongst others we find them united. We are therefore at liberty to make conjectures as to which state of things is the earlier ; and this paper has been in effect an attempt to show that the union may reasonably be considered to have preceded the divorce. So far as that conclusion has any probability, it may encourage us to enquire whether there are any other institutions which, though they appear to be isolated from one another in the present condition of the Australian tribes, may originally have been united. Now, the institution which is most regularly present in religions of all kinds is Sacrifice ; and there are certain rites, or at any rate practices, observed in Australia which have generally been considered to be a primitive form of sacrifice and the sacramental meal. These practices form a part of the system of totemism. A totem is in nearly all cases an edible plant or animal, after which the totem-tribe is named. The animal or plant is regarded with respect or reverence by the tribe whose totem it is ; and when the season for eating it arrives, there must be a ceremonial eating of it by the Head Man of the tribe to whom it is a totem, before men of other tribes will eat freely of it. This custom has obvious analogies with the fact that most peoples, in a more advanced stage of social and religious development than the Australians have reached, will not eat of the kindly fruits of the earth until an offering of the first - fruits has been made to the gods. Evidently amongst these more advanced peoples it is not felt to be safe or proper to eat of the harvest until a rite has been performed which is of a religious character and significance. The Australian tribes also feel that it is not safe or proper to eat until a certain ceremony has been performed. And it was an easy conjecture that as the ceremony was in the former case religious, so it must be in the case of the Australian tribes. But though the investigations of Messrs Spencer and Gillen have shown that the ceremonies are of a very elaborate character, they have not shown them to be possessed of any religious character. The Religion, if any, of the black fellows is to be found not in these Intichiuma ceremonies, but in the Initiation ceremonies to which I have already so often alluded. It seems, therefore, that if we take up the religious rite of Sacrifice and trace back its history, we find when we get back to its earliest and most rudimentary form that there is nothing religious in it. Indeed, Mr Frazer and Messrs Spencer and Gillen, independently of each other, arrived at the conclusion that the Intichiuma ceremonies were magical in intent, and designed to secure by magical means a proper supply of food.

RELIGION IN EVOLUTION

I must now call to your minds once more
the fact that the Northern tribes of Central
Australia described by Messrs Spencer and
Gillen and the South-Eastern tribes described
by Mr Howitt differ in many important
respects; and I must add to the other differ-
ences the following : it is, that the Intichiuma
ceremonies, which are regarded by Mr Frazer
and Messrs Spencer and Gillen as magical in
intent, are found in those Northern tribes which
have ceased to believe in the All-father ; and
they are not found in the South-Eastern tribes
who continue to believe in "the father of all
of us." Now, I have already put forward the
supposition that the South-Eastern belief in
"the father of all of us" is earlier than the
Northern tenet that the All-father is a mere
bogey; and if we must conjecture why there
are Intichiuma ceremonies in the one set of
tribes and not in the other, I would suggest
that belief in magic tends to flourish at the
expense of Religion. Where the belief in
"our father" continued operative, the magic
which was developed in the Intichiuma ceremonies did not flourish. Where the religious
belief declined, and because it declined, the

practice of magic grew. My next point is that sacrifices, which are a part of the ritual of Religion, are frequently borrowed by magic and used for magical purposes. And I suggest with regard to the Intichiuma ceremonies that if, as they are now practised, the religious element is wanting and the magical element is predominant, it is because the religious element has evaporated from them, and not because it was never there. It has evaporated from them, as it has evaporated, in the case of the Northern tribes, from the Initiation ceremonies. As we conjecture that religious belief was once present in the Northern tribes' Initiation ceremonies, though traces and survivals of it are now alone to be found, so we may conjecture that it was originally at the root of their Intichiuma ceremonies.

We may say, then, that the history of the institution of Sacrifice leads us to expect to find an early form of its development in Australia. What we find is an institution which would be sacrificial if only it were religious. We may, if we like, stop at that point. If we do, then we have an instance in which a cardinal feature of Religion has been evolved

37

from antecedent phenomena of a non-religious kind ; and then we must bear in mind the fact that those phenomena *ex hypothesi* do not, and did not, constitute Religion. If we do not feel inclined to rest content with a theory that requires us to suppose that Religion has borrowed from magic the conception and the mechanism of the sacramental meal, then we may scrutinise the Intichiuma ceremonies in the hope of conjecturing their antecedents and their true relation to the other social and religious institutions of the Australian tribes. We may see in the Intichiuma the same process of religious degradation as we suppose we see in the Initiation ceremonies of the Northern tribes. We may conjecture that the ceremonial eating of the totem animal or plant, which at the present day appears magical in intent, was originally in the nature of a sacrifice and a sacramental meal ; and that the same tendency which amongst the Northern tribes robbed the Initiation ceremonies of their religious value, also emptied the Intichiuma ceremonies of their religious content. Among the South-Eastern tribes, on the other hand, the religious intent of the Initiation ceremonies either still survives or may be confidently traced ; but survivals of the sacramental meal have not been noticed. It would not be safe to say that because they have not been detected that they have not existed, or that they do not exist amongst the tribes whose manners and beliefs have not yet been examined. It would have been easy to deny that the Northern tribes had any belief in the All-father, had not the belief been discovered among the Kaitish ; and even so Messrs Spencer and Gillen were led to minimise its value. But even if we assume that survivals of the sacrificial meal do not now exist amongst the South-Eastern tribes, it is not unreasonable to regard the solemn eating of the totem amongst the Northern tribes as such a survival. We may assume that the Northern and the South-Eastern tribes are descendants of common ancestors, and that the social and religious institutions of the descendants have been evolved out of those of their ancestors. For instance, the totemism of the present tribes would generally be allowed to have descended to them from very early times— how early it is impossible to guess, but two facts seem to relegate it to a far-distant past. The first is that the origin of totemism is—unless light is thrown upon it in a forthcoming work by Mr Andrew Lang—quite lost to view, and does not even lend itself to plausible conjecture. The next is that the native tribes must have been in Australia not merely for some centuries, but for a vast number of centuries. They have not remained during that period of unknown length unchanging and unchanged. Primitive indeed they still are, but not primeval. They have gone through a long, though probably not a rapid, course of evolution. The totemic system of the South-Eastern tribes, viewed as a system regulating kinship and marriage, is far more highly evolved than that of the Northern tribes, and therefore more remote from the system of their ancestors. The fact, therefore—if it is a fact—that no survival of the sacrifice of the totem animal is to be found among them, is the less to be wondered at. On the other hand, it is well to remember that, if totemism has survived conspicuously in the rites of the Intichiuma, it also survives in the teaching given to boys at the Initiation ceremonies:

the boys are there taught that "they must on no account reveal any of the secret matters of the totems to the women and children" (S. and G., *l.c.*, p. 503). The implication obviously is that both ceremonies, both the Intichiuma and the Initiation rites, are descended from a ritual in which the doctrine taught was belief in the All-father, and in which the rites observed consisted in a sacrifice or a sacramental meal.

If totemism were wholly absent from either the Intichiuma or the Initiation ceremonies, there might be no reason for casting back for some system of ritual and belief from which both may be believed to have been evolved. If the suggestion that the original purport of the Intichiuma ceremonies was purely to provide by magical means a proper supply of food, were unanimously adopted by those qualified to judge, it might be well to set aside straightway any other theory on the subject. But Mr Howitt hesitates to endorse the suggestion, and hesitates on the ground that "the origin of totems and totemism must have been in so early a stage of man's social development that traces of its original structure cannot be expected in tribes which have long passed out of the early conditions of matriarchal times " (p. 151). I will conclude this lecture, therefore, by saying that if the origin of totems and totemism must have been in so early a stage of man's social development that traces of its original structure cannot be expected to be found in the Australian tribes, then perhaps we cannot expect to find among them the origin of Religion either.

II

IN this lecture I shall deal with the Science
of Religion and with Evolution. I shall
point out first that the Science of Religion,
like any other science, is based upon and starts
from facts of experience; and next that the
facts from which science generally, and therefore the Science of Religion, proceeds, are not
facts of merely individual experience, but of
the common experience of mankind. I shall
then note that science is not the facts with
which it deals and to which it relates, but is an
abstraction from them. Next, the Science of
Religion, like all other sciences, abstracts
from, that is to say ignores deliberately, the
Freedom of the Will, or at least the possibility
that the Will is free. Finally, I shall argue
that the theory of Evolution, like the science
on which it is based, is an abstraction: it deals
with growth, with the process of Evolution.
And growth and process are abstractions:
they are ways in which Reality appears to us

or may be conceived to present itself to us. But they are Appearances ; and in the case of Religion, at any rate, they may be distinguished from the Reality with which the soul is in communion when it lifts itself to God, or strives and yearns to cling to Him. In fine, Science of Religion is something very different from Religion, and the theory of the Evolution of Religion is not a religious experience.

The outlines of this lecture having been thus given, we may proceed to a consideration of the first step in the argument, viz. that the Science of Religion, like any other science, is based upon and starts from facts of experience. Now, it would appear that the facts upon which the Science of Religion is based must be facts of religious experience, just as those on which any physical science is based must be material facts : Religion is as essential to the Science of Religion as matter is to the physical sciences. But neither clause of this last sentence will command unanimous assent : those who regard Religion as an illusion cannot agree that a Science of Religion is possible only on the assumption that Religion is real and true ; and the reality of matter is similarly questioned even by some men of science. I propose, therefore, to treat of both questions— that of the existence of matter and that of the existence of Religion—and to treat of them separately.

I will begin with the question of matter. About the existence of matter, of things material, the ordinary non-metaphysical, non-scientific mind has no doubt : matter and material things do exist. And science, which starts from the facts of ordinary experience and from the position of common sense, has no doubt either about the existence of matter and things material. It is metaphysic and metaphysicians — or, rather, some meta-physicians—who deny or doubt the existence of matter. And it is generally admitted that the doubts which metaphysic raises, metaphysic must settle. The business of science, on the other hand, is to keep clear of metaphysics and metaphysical problems : it has to ascertain all that can be ascertained about the co-existence and interaction of things material and about the laws of causation which express and explain their

45

relation and succession. But it is no part of the business of science to enquire whether, or in what sense, matter and things material exist. Those are questions for metaphysic to discuss—and to settle if it can.

Then, what is the position of physical science in the meantime? It is hung in air, so to speak. And, supposing that metaphysic came to be in a position to demonstrate to any ordinary person who chose to listen to the demonstration that matter does not exist, would physical science then collapse? For it would be shown that physical science is based on an assumption, viz. that matter exists, and that the assumption is patently wrong. Indeed, it is not necessary for my present purpose to ask what would be the position of physical science if metaphysic demonstrated undeniably the non-existence of matter. It is enough to point out that the question of the existence of matter is discussed by metaphysic ; and the mere discussion is quite enough to show that for the metaphysician, at any rate, the existence of matter is not as certain as it is for the ordinary mind and for many men of science.

RELIGION IN EVOLUTION

The position then seems to be that physical science postulates the existence of matter and things material, and is based upon that postulate. Withdraw that basis, show that that postulate is one which cannot reasonably be granted, and apparently the physical science which is built upon it must collapse. It is not surprising, then, or unnatural, that men of science should look upon metaphysic with some degree of impatience, suspicion, or contempt; for they find themselves attacked by a weapon against which science is incapable of defending them. And that is a position which is eminently unsatisfactory for those who hold that what is not science is not knowledge. The only thing for those to do who hold that view is to shrug their shoulders at metaphysic and say, "Everybody knows that science is not hung in air, is not a baseless vision : therefore matter does exist, and if metaphysic pretends it does not, so much the worse for metaphysic." Having delivered his soul thus, the man of science may go back to his science, somewhat ruffled perhaps, but not the less satisfied with himself. We, however, who are left behind pondering, must see

47

whether his case could not be rather better stated than he himself has put it.

What we want to do is to place physical science on a basis somewhat more satisfactory to metaphysic and somewhat safer for science than is afforded by the assumption that matter exists and that science is based upon that somewhat ambiguous assumption. If we come to reflect upon it, what science is built upon is experience — the experience which the man of science who has made an experiment or a discovery has himself gone through, and which any other person who chooses may equally experience. What the scientific discoverer asserts is that, under given circumstances, anyone may have the same experience, get the same results from the experiment, as he had. Now, if this be so, there seems to be no reason why matter should ever be dragged into the question. There is no reason why we should go beyond the statement that such has been, and, under the same circumstances, will always be, the experience of any man who chooses to go through the experience. The experience of knocking one's head against a brick wall is not in the least affected or modified by any view we may hold as to the existence or non-existence of matter. The experience may or may not be rendered more intelligible by the metaphysical assumption that matter exists; but the experience comes first, the assumption comes afterwards—and the experience remains equally valid, even if the assumption never follows, or does follow and is subsequently shown to be an untenable assumption.

If, then, we may now take it that physical science is built upon experience, and not upon any such dubious assumption as that matter exists, we may perhaps venture to suggest that the Science of Religion rests upon the same foundation as any other science, viz. upon the foundation of experience; and assumes, like every other science, that the experience on which it is based is a real experience. Here, however, in this last assumption we touch upon a point of fundamental importance for the Science of Religion—of fundamental importance because it raises the question whether the object of the Science of Religion is to enquire whether the subject which it investigates really exists. We may perhaps best answer the question by the familiar device of asking another question, viz. whether it is the business of the physical sciences to enquire whether matter exists. It is true indeed that the question whether matter exists is a metaphysical, not a scientific question—a question which was discussed by Bishop Berkeley, and will not be solved on scientific considerations, such as those that are sometimes advanced. It is true also, therefore, that a man of science may, like Professor Huxley, be no Materialist and may hold to Berkeley's view. It might therefore be argued that the man of science may quite well engage in the study of any of the physical sciences without pledging himself or indeed holding any opinion whether matter does or does not exist. And from this point of view it might be held that the student of the Science of Religion is also equally free to pursue his science whether he believes Religion to be or not to be in any sense real or valid, or even without holding any opinion on the subject one way or the other.

But this mode of argument will on reflection prove hardly tenable. Whether matter does or does not exist is indeed a question of metaphysical speculation; but the reality of the experiences through which the student of science goes in his experiments and his observations must be admitted from the very beginning, or else the foundation of science is removed and the superstructure collapses. What is thus true of physical science is also true of the Science of Religion; unless the reality of religious experience be a fact undoubted from the beginning, the Science of Religion has no basis to rest upon, and collapses in consequence.

If then the Science of Religion, like any other science, is based upon and starts from facts of experience, we may now proceed to our next point, which is that the facts from which science generally, and therefore the Science of Religion, proceeds, are not facts of merely individual experience, but of the common experience of mankind. This proposition, however, true though it be, is by no means universally admitted to be true. Amongst those who would deny it are many profoundly religious minds: they claim that no one shall or can stand between a man and his Maker, and that real Religion resolves

itself ultimately and exclusively into the relation in which a man's soul stands to his
God. So strongly is the truth contained in
these propositions emphasised by some minds,
that they overlook practically altogether the
fact that no individual man is, or ever can
be, independent of the religious experience of
those with whom he is in sympathy. They
ignore in their theory, though not in their
practice, the fact that every one of us depends
on the spiritual experience of others, and
learns from them what he might otherwise
have remained in ignorance of. Not only
may he learn what to seek : he may learn
what to shun, for he may require to be taught
how to pray and give thanks, and to be taught
how the Pharisee's thanksgiving differs from
the Publican's prayer.

If anyone will read Professor James's
" Varieties of Religious Experience," he will
there find countless instances of the consequences which ensue when the individual
soul adventures forth into the spiritual world
alone, without guidance. As you read the
records which he quotes of the experiences of
solitary souls, the region of prayer and

spiritual expansion seems a realm of individual extravagance, of disordered visions, of
spiritual hallucination. The conclusions which
may be drawn from this record will differ
according to the different pre-suppositions
with which it may be read. If we start with
the pre-supposition that Religion and religious
experience is a purely individual affair—that
the ultimate and only basis for Religion is
what I myself experience—then there are
two alternatives before us. Those alternatives are either to believe or not to believe
that there is something valid and real in
religious experience. If there is something
valid and real, then the question arises
whether all these experiences are alike valid,
real, and religious. To many or most of those
who have had these experiences—probably to
all of those who have recorded them—they
appear to be undoubtedly and all equally
alike real and true. We may then, if we will,
take up the position that what appears to one
individual real and true is real and true for
him ; and is neither real nor true for any
other individual who happens to differ from
him. In a word, there is no absolute truth

or reality whatever : you assert that A is A ; I deny it, and both the assertion and the denial are in the same sense true and in the same sense false. If the individual or individual experience is the final judge, beyond whom lies no court of appeal, then the spiritual extravagances quoted so copiously by Professor James are in the final resort just as valid, true, and religious as the experience of the founder of any of the higher Religions—or of the highest.

Now, that is exactly the position taken up by those who accept the other of the two alternatives already mentioned, and who hold that there is nothing valid or real in religious experience : all religious experience alike is invalid—it may differ in its manifestations— the forms folly may take are innumerable and incalculable—but the one thing certain is that it is a purely individual affair, and that not all individuals have it. Doubtless the fact that they themselves have it not, is the proof conclusive to them that Religion is a purely individual matter. But they are not, and rightly are not, content to leave it an open question. If it is an open question, then the man who believes has just as much right to do so, and is just as likely to be right in doing so, as the man who does not. And so long as the matter is left there, there is always the possibility that want of Religion may after all be an abnormal condition—as abnormal, for instance, as any of the spiritual extravagances quoted by Professor James. In fact, the occurrence of a certain small percentage of non-religious minds would no more prove the non-existence of Religion, than the occurrence of a small percentage of colour-blind persons in the population proves that the rest of us have no experience of colour and are mistaken in imagining that we have. The realm of music and the world of art can scarcely be pronounced illusions in order to gratify the tone-deaf or colour-blind, who cannot believe, or wish not to believe, in the existence of what they cannot appreciate.

But is the want of Religion an abnormal condition? If we may take it that Religion involves a belief in a personal God, then, as Mr M'Taggart points out in his "Hegelian Cosmology" (p. 74), "mankind has been by no means unanimous in demanding a personal

RELIGION IN EVOLUTION

God. Neither Brahmanism nor Buddhism makes the Supreme Being personal . . . and in the Western world many wise men have been both virtuous and happy who denied the personality of God"; for instance, Hume, as Mr M'Taggart points out. Not only have there been in Europe "cases of men of illustrious virtue who have rejected the doctrine of a personal God," but the number of such cases is, Mr M'Taggart suggests, increasing. He says : " Whether the belief in a personal God is now more or less universal than it has been in the centuries which have passed since the Renaissance cannot, of course, be determined with any exactness. But such slight evidence as we have seems to point to the conclusion that those who deny it were never so numerous as at present." Let us then accept Mr M'Taggart's conclusion, and let us draw the inference that the number of those who disbelieve will go on steadily increasing until the proportions have been reversed and those who believe are in as small a minority as are those who disbelieve now. Will the fact that the majority has shifted make any difference to the merits of the question ? At the present moment those who are in the minority are quite satisfied that the majority are wrong: will the case be any different with the minority of the future? If Religion and religious experience are a purely individual matter, and each man's experience or want of experience is final for him, and there is no appeal beyond him—if the individual is indeed the measure of all things, then it is irrelevant for him to enquire or consider what other people think: the matter is decided for him by his own experience.

That people do differ in this way is matter of fact. If we enquire why they differ thus, the answer probably is that given by Hegel in his "Philosophy of Religion" (i. p. 5), viz. because the will is free. It lies with the individual, because his will is free, to accept or reject Religion. We are probably never so distinctly conscious of the freedom of the will as we are when we definitely decide to reject it or to accept it. That decision is indeed a purely individual affair—a matter of purely individual responsibility. But the worth of the decision and the value of the grounds on which it is reached are not determined thus.

RELIGION IN EVOLUTION

If the individual is assumed to be the measure of all things, then he who decides to accept Religion is just as right as the man who decides to reject it—or, if we like to put it so, the one is just as wrong as the other. But no man who holds a strong opinion on this point—whether he believes or disbelieves—can reconcile himself to this conclusion. If he is in the minority now, he consoles himself with the thought that eventually, indeed perhaps even now, all really enlightened persons will be found on his side. If he is in the majority, he has no difficulty in believing that either some people are colour-blind and tone-deaf in this respect, or that the matter is one in which nobody is congenitally incapable either of Religion or of atheism, and everybody may freely will to accept or reject either. But that those are right who decide with him, or whose decision he concurs in, the man who feels strongly and wills decidedly has practically no doubt. That is to say, his belief is normal and is really right: the opposite is abnormal and is ultimately wrong. Just as in physical science the accepted ground is the experience which is open to all enquirers alike, and those conclusions are valid which are confirmed by the experience of all who choose to put them to the test; so in the Science of Religion the only solid and scientific basis is the experience which every man may consult if he will. It is this experience from which the Science of Religion starts, and to which it returns—the experience in which the individual partakes, but of which he is not the sole possessor. The starting-point is not my individual experience, or my interpretation of my experience, in the Science of Religion any more than in any other science. In every science alike the basis is the fact that a given experiment can be made or assertion proved by the experience of every individual. Unless there is this community of experience, there is no science; truth is that which is true for everybody. Objective truth is that which is true not because a man thinks it so, but whether a man "thinks it so or not, and which must be judged to be so by all rational beings"; because "all rational beings, in so far as they judge rationally, must necessarily judge similarly of the same matter." It is not the experience of any one individual on which science is based. If it were, there could be no science, for science assumes that it is dealing with facts ultimately verifiable in the experience of any man capable of reading, and willing to read, his experience aright. But the facts from which the Science of Religion starts are not facts of merely individual experience, but of the common experience of mankind. Hence it is that public worship is in all countries and in all ages felt to be an essential condition of Religion. The congregation of worshippers is a spiritual community, and without this spiritual unity there is no Religion.

It now becomes necessary to note that no science is the facts with which it deals and to which it relates : every science, and therefore the Science of Religion, is an abstraction from the facts. First, then, what are the facts with which the Science of Religion has to deal? Next, in what sense is the Science of Religion an abstraction from them? And finally, with what object is the abstraction made?

The late Professor Sidgwick, in his "Methods of Ethics" (Book III. ch. i. § 4), insisted that the existence of morality should be discussed quite independently of the origin of morals ; and that the question of the validity of ethics was quite independent of and could not be affected by any conclusions which might be reached as to the manner in which morality as a matter of historic fact originated. The principles which he laid down as to the proper method of discussing the existence, origin, and validity of morals are equally applicable to the existence, origin, and validity of Religion ; and I shall now proceed to apply them.

The first question, as to the existence of morality, can, he says, "only be determined by introspection, together with the observation of the present phenomena of other minds"; and what he says as to the method of determining the existence of morality obviously applies with equal force as to the method of determining the existence of Religion. The assumptions made both with regard to Religion and with regard to morality are, first, that the phenomena are exhibited generally in other minds ; and next, that it is possible to observe "the present phenomena of other minds." Neither Religion nor morality is confined to this mind or that, but is to be found actually or potentially in all minds ; and both Religion and morality imply that "the present phenomena of other minds " are accessible to us, and that when we have gained access to them we find that their experience is actually or potentially ours. Without such spiritual communion there would be neither morality nor Religion. The experience in which we participate is yours or mine so far as we choose to partake in it, but it does not cease to exist if or when we choose to turn aside from it.

Before, however, we can leave the matter of the existence of Religion to turn to the question of its origin, it is necessary to define it. What Professor Sidgwick said of morality applies with equal force to Religion : " It seems premature to enquire into the origin of anything before we have ascertained what it is." This statement of Professor Sidgwick's may then be supplemented by the obvious comment that if we have ascertained what a thing is we are in a position to state what it is, that is to say, to define it. And till the student of the Science of Religion has some idea what Religion is, he will not be able to recognise it when he sees it, and will not advance the cause of his science. Now, if there is to be Religion, there must be, as we have seen, a body of individuals : they must have a common purpose, each must be conscious of that common purpose, and the congregation must be so far, and in that sense, in spiritual communion. But all these conditions are equally requisite and equally realised whenever any body of men work together for any purpose. The conditions may be indispensable to Religion, but they may be realised without resulting in Religion. The one thing wanting from them is the one thing necessary to Religion, viz. the sense which the worshippers have that they are in spiritual communion not only with each other, but with their God —and that God conceived not merely as a "principle of unity," but as a personal God. But this sense is not merely a cold intellectual perception of a fact which arouses no particular emotion : it is a sense of love—of love towards one's God and towards one's neighbour.

These, then, are the facts, as disclosed by introspection of one's own mind and observation of the present phenomena of other minds, with which the Science of Religion has to deal ; and the facts obviously are something different from the Science which deals with them. We may say that the facts are concrete facts of experience, and that Science is abstract, or deals in abstractions made from the facts; and if we say so, we are making a statement which is undoubtedly true, but which is liable to misinterpretation. It would be misinterpreted if it were understood to imply that Religion is experience and science is not. The man of science in conducting his experiments or drawing his conclusions is certainly undergoing experience—experience as direct as it is possible to have. But science, say Science of Religion, which is itself an experience, is not experience of the religion which it dissects—nor are the dissected members the religion in which they were elements. A man may, for the purposes of science, study a religion which is not his own, and in so doing his experience is plainly different from that of a believer when practising his religion. For the purposes of science a man may study the religion which is his own; but so far as he treats it scientifically his attitude is quite different from that in which, as a believer in it, he stands towards it: the experience through which he goes is different in the two cases. The difference in the two cases is not merely that his attitude is different but that with which he is dealing is different. As a man of science he takes the religion with which he is dealing in abstraction: he abstracts from it, and sets aside, for one thing, the religious feeling or emotion which is the very breath of its being, and without which it is indeed fit for the dissecting table, but is no longer the religion which animates and vivifies those for whom it is a living thing and the vital truth. That is why Science of Religion is — not altogether unjustifiably — to some minds so repellent. The man of science may "peep and botanise upon his mother's grave," but to do so he must for the moment banish from his mind the relation in which he stands to it : the turf must for the moment be as any other piece of turf : it must be taken in abstraction from its other relations. It is with such abstractions that Science of Religion deals, and only with such abstractions. If it is with his own religion that the student is concerned, it requires no great effort to realise that he is then dealing with an abstraction. If it is with a religion not his own, he may easily forget that there are, or have been, those for whom it was no abstraction, no *caput mortuum*, but something very different from that which he has before him. If the student has no religion of his own, he may easily fall, and in some cases undoubtedly does fall, into the fallacy of imagining that for no one was or is religion anything but the unreal thing which it is for him. In any case it is clear that the student of the Science of Religion cannot believe all the religions which he studies, and that any religion taken apart from belief in it is an abstraction. The Science of Religion therefore deals with an abstraction from the facts, and is not the facts with which it deals and to which it relates.

If now we enquire with what object this abstraction from experience is made, we must reply in the first place that the Science of Religion is a historical science, and as such its object is to trace the Evolution of Religion. Whether, when the evolution of religion has been traced, all that science and philosophy can do for religion has been done, is a point to which we shall have to return hereafter. For the present we may be content to note that any theory of the evolution of religion must concern itself, amongst other things, with the question of the origin of religion; and following the example of Professor Sidgwick, who, in dealing with morals, insisted on the necessity of sharply distinguishing between the origin and the validity of morality, we shall draw the same distinction in the case of religion. In words which apply to religion as well as to morality he said: "It seems to be frequently assumed, that if it can be shown how certain mental phenomena, thoughts or feelings, have grown up—if we can point to the antecedent phenomena, of which they are the natural consequents—then suddenly the phenomena which we began by investigating have vanished: they are no longer there, but something else which we have mistaken for them, viz., the 'elements,' of which they are said to be 'composed.'" Thus, to apply to religion the argument which Professor Sidgwick used of morality, if the Science of Religion can point to the antecedent phenomena of which religion is the natural consequent, then it is

RELIGION IN EVOLUTION

sometimes supposed that religion is thereby exploded: it is no longer there, but only the elements—say fear or magic—of which it is supposed to be composed. This is a fallacy, it is hardly necessary to say, into which they are particularly prone to fall who hold that there is nothing "in" religion: they trace its origin to certain antecedent phenomena, and believe that those phenomena—fear or the belief in magic — are religion, and that it is only by a mistake that religion is ever considered to be anything else. But as Professor Sidgwick noted, the laws of belief are not like the laws of chemistry; or, in his own words: "The psychical consequent is in no respect exactly similar to its antecedents, nor can it be resolved into them: and there is nothing, at least according to the ordinary empirical view of causation, which should lead us to regard the latter as really constituting the former." That is to say, religion regarded, as for the purposes of the theory of Evolution it must be regarded, as a psychical consequent which ensued upon certain antecedents, "is in no respect exactly similar to those antecedents" — say fear or the belief in magic—"nor can it be resolved into them": they do not "really constitute religion."

The one thing necessary for the theory of Evolution is that it should be free. The one thing necessary if its results are to be accepted by religious minds is that it should be matter of common knowledge that, whatever view of the origin of religion may be taken, its validity remains unaffected. Professor Sidgwick said, "It has been very commonly assumed on the one side that if our moral faculty can be shown to be 'derived' or 'developed,' suspicion is thereby thrown upon its trustworthiness: while on the other hand if it can be shown to be 'original' its trustworthiness is thereby established. The two assumptions seem to me to be equally devoid of foundation." The same view is taken by Professor Sorley, the successor to Professor Sidgwick's chair: he says in his "Ethics of Naturalism" (2nd ed., p. 133): "It cannot be held that moral intuitions are invalid because evolved. The evolutionist will certainly go very far wrong,

as Mr Sidgwick points out, if he maintains that 'a general demonstration of the derivedness or developedness of our moral faculty is an adequate ground for distrusting it.'"
It is scarcely necessary for us to insist that what is thus said, and repeated, of morality is equally true of religion : its trustworthiness, its validity, is a question quite apart from the question of its origin.

The Science of Religion then, as science, is not concerned with the question of the validity of religion. Indeed science, generally, has not to do with the question whether the experience on which it is based is or is not trustworthy : it takes experience for its basis and as the test of the trustworthiness of its conclusions. It leaves to metaphysic the enquiry whether its basis and foundations are sound. Further, each particular science limits itself to the investigation of some particular aspect or department of experience ; and takes that department apart from — in abstraction from — the rest. The question whether some given religion is or is not valid is a question with which the Science of Religion has not to do : it takes any religion, with which it deals, apart from the question whether it deserves belief, and deals with it in the abstract. That, of course, is not the only abstraction which the Science of Religion makes from concrete religion. Another, and one with which we will now go on to deal, is that it abstracts from the Freedom of the Will. It is no part of my intention to prove that the will is free. I assume that it is so, and that science—quite legitimately for its own purposes—sets aside the assumption. I only wish to point out that science does not begin by disproving the freedom of the will: it begins by assuming that there is no freedom. Consequently nothing in science can prove that the will is not free. When all the deductions have been drawn that science is capable of drawing, the question whether the will is free remains untouched. Science assumes that there is no freedom of the will, and the fact that the conclusions of science are in harmony with its original assumption no more proves the assumption to be true, than the coherency of Euclid proves that two straight lines cannot enclose a space.

There is, of course, nothing arbitrary in the

proceedings of science, when it decides to take the facts of experience in abstraction from the freedom of the will, with which in experience they are or appear to be associated. There is nothing arbitrary, for the simple reason that the object of science is to discover the causes of things and the laws according to which things must be supposed to happen if we are to have any scientific knowledge of them. If, then, our object is to discover the laws and causes of things, we must assume that everything has a cause, that nothing can happen without a cause, and that a cause can only produce a given effect. This assumption is fundamental for all science, and consequently it is fundamental for the Science of Religion. For science or by science the whole process of religion must be regarded as the necessary outcome of laws and causes which could not be otherwise than as they are : the whole process is studied apart, in abstraction, from the freedom of the will — the individual is supposed not to be free in his actions, his beliefs, his aspirations, or his want of belief and his turning away from the things of religion. The Science of Religion is abstract and deals with abstractions — with certain aspects of experience —just because its object is to ascertain and state laws, which laws are themselves abstractions.

Science, then, is an interpretation of experience; but the interpretation of a thing is not the thing interpreted; nor does the original text disappear and cease to exist, because a translation of it appears, say, in Bohn's series. The translation doubtless helps the student to a better understanding of the text; but it is not the text which it interprets—more or less inadequately. The translation is neither the original nor is it final. The original text is there before the translation is made; and it is there after the translation is made. And however good the translation is, it is a translation to the end; and its object and justification is not to take the place of the text or to pose as being the original, but to help us to a better understanding of the original.

Now science is a translation of the original text of experience : it translates that experience into laws and causes; it puts it into a shape and gives it an appearance other than its own. And it does so in order that we may go back to the original, the real thing, better fitted to appreciate it. But it cannot be denied that in the present day, the age of science, the general notion is that we must stick to the translation, and that there simply is nothing else to go to or go back to. The current notion is that the translation is the original text, that science is not a means but an end, that when we have read the scientific translation we are entitled to deny that there is any original text, and to assert that science is the final truth, not merely about the abstractions with which science deals, but about the experience from which they are abstracted.

In one respect, indeed, common-sense does feel that there is some discrepancy between the experience on which science is based and the science which is built upon it—and that is the matter of Free-will. But as common-sense is unwilling to part with either, it retains both, without any attempt to reconcile them. Yet the Universality of Causation which science postulates is incompatible with the freedom of the will which common-sense recognises. If we start by assuming the freedom of the will, we must at least deny that the law of

cause and effect holds good of everything—
we must infringe to some extent upon the
universality of causation. We may perhaps
imagine that we shall be able to pause satisfactorily and permanently, if we draw a
distinction between mind and matter, and
regard the one as the abode of spiritual freedom and the other as the region in which
causation is universal and nature is uniform.
But in that case either mind and matter
interact upon one another or they do not.
If they do interact upon one another, then
mind, in so far as it is thus acted upon, is
so far subject to the law of causation ; and
matter, so far as it is thus acted upon, is no
longer subject to the law of cause and effect.
Yet, how can we imagine or believe material
things to be set in motion or deflected in their
motion except by material things ? or how can
material things impinge upon spiritual beings ?
Whence comes the motion in the one case and
what becomes of it in the other ? And how is
either theory compatible with the Conservation
of Energy ? in either case the sum total of
energy must be a varying amount. Feeling
these difficulties, we may incline to the view of

those who have held that there is no interaction between mind and matter, but that consciousness is epiphenomenal, that is to say, it accompanies or is concomitant with movements and changes of matter, as a shadow may accompany the locomotive which casts it. But in that case, we who believe in free-will get no help, for it is the material train which casts the shadow—the epiphenomenal consciousness, that is to say, the shadow, does not move the train.

Thus the difficulties in which we are involved, if we draw a distinction between mind and matter, and imagine that distinction to be ultimate and fundamental, are from any point of view great; and they become intolerable, if we are in earnest with the belief that God is a spirit, and that the basis and reality of all things is spiritual. In fine, if the spiritual is the real and is the only reality, then the universality of causation and the uniformity of Nature can only be aspects of the real, ways of looking at it, abstract conceptions of it—appearances. To this point of view science itself, when it passes into evolution, seems to be tending. The law of causation is that under the same circumstances the same result will ensue—if identically the same antecedents recur, the same consequence will follow. The only question is a question of fact, viz., whether identically the same antecedents ever do recur. If they did, the course of the world would repeat itself as if it were a recurring decimal. But from the point of view of evolution the same thing never does recur : each stage in that evolution is different from any that preceded and from any that will follow. The uniformity of Nature, in this sense of the words, is abandoned : we continue to assert strenuously that if or whenever the same cause recurs the same effect will follow, but, when we are promulgating the theory of evolution, we maintain that the same cause never does recur. And certainly, in our own personal experience, the same circumstances never are repeated : at no two periods of our lives are the circumstances, however similar they may be, exactly the same—and, if they were, we at any rate are not the same.

The question, however, now suggests itself whether we have got rid of the law of causation, because we have pointed out that there is no room in the theory of evolution—or the process of evolution—for recurring causes. We can scarcely believe that we have. A cause is no less a cause, even if it has never happened and never produced its effect before, or will never occur and therefore never produce its effect again. We have not got rid of causation because we have got rid of the uniformity of causation. Every single cause that acts is a cause, even if no two causes in the whole course of the universe's evolution are the same. And in a universe which evolved in this way, there would be no event uncaused and no room anywhere for any free-will. Universality of causation is incompatible with freedom of the will; and the theory of evolution abstracts from the freedom of the will, it is built upon the assumption that no stage of evolution could have been otherwise than it was.

Thus the theory of evolution is essentially abstract, in that it sets aside the freedom of the will. It is also abstract in another sense, viz., that it concentrates itself on one aspect of things—their growth and development. That

78

is to say, it accepts without question the reality
of Time and Space. It is abstract again in
yet another sense, viz., that it investigates the
process of evolution in time and space, without reference to—without prejudging—the
question whether there is a God. The theory
of evolution then is abstract through and
through. It starts from experience, but it
confines itself to certain aspects of experience.
Eventually, therefore, we must face the
question whether a theory which avowedly
confines itself to certain aspects of experience,
can be accepted as a satisfactory explanation
of the whole of experience. The only ground
on which we could so accept it would be that
we had reason—quite apart from the theory of
evolution—to believe that the will is not
free, that time and space are realities, and
that the process of evolution requires no God.
If on the other hand, we hold that the will is
free, or that time and space are not ultimate
realities, and that there is a God, then the
theory of evolution will be for us not Reality
but one aspect or Appearance of Reality. It
will enable us to understand the reality of experience in some respects and from some

points of view the better, as any translation may help us to a better understanding of the original. But like all translations it is inadequate and even in some respects misleading. The question then arises whether time and space are real, for if they are only appearance, the process of evolution also is not Reality but an appearance given to Reality. This question will occupy us in the next lecture.

III

IN this lecture I propose to discuss the subject of Time and Space. I wish to show that Time and Space are ways in which we interpret experience—ways in which we dissect experience. And I use the word "dissect" advisedly, as wishing to imply that experience must be dead before we can lay it out in Time and Space. In the first place, it is when we *reflect* upon experience that we arrange it in Time and Space, not when we are aware of it; and, next, any experience is, at the moment when we have it, a live experience, so to speak, whereas, when we "re-flect" or turn back upon it, it is, as it were, dead; and then we lay out its corpse in Time and in Space. My object therefore is to argue that in our living experience we have to do with the timeless and the non-spatial. Of course, the common-sense notion is that Time is a reality, that things take place and are in Time. That notion of course involves

F

the distinction of past, present and future as something given to begin with, and not as a distinction introduced by us. In the same way the popular idea is that a cause is something given to start with as distinct from the effect and as really separate from the effect. It is, however, clear on the least reflection that the popular idea is quite untenable : the distinction is one which we put upon the facts and by which we interpret the facts—but it is not in the facts. For instance take the case of the explosion of a barrel of gunpowder : enumerate all those conditions which are necessary to the production of the effect, *i.e.* without which the effect would not take place : the gunpowder must be there, in a confined space, it must be dry, it must be in an atmosphere which permits of explosions, etc., etc., and a light must be applied. Now the cause is the sum of conditions necessary to the effect. Unless and until all the conditions are there, the cause does not exist. But the moment the conditions are all there, the effect is produced. *We* may distinguish in words between the cause and the effect, but the distinction is a verbal one, introduced by us, and not in the facts. We may enumerate all the conditions of the explosion—spark included—and having done so we can pause, or, without pausing, we can say " and then the explosion followed." Indeed we actually believe it, for that is what we read in the newspaper. But though we pause and distinguish between the effect and the sum of conditions necessary to produce it, there is no such pause in actual fact, for the sum of conditions is not only necessary to produce the effect, it constitutes the effect. The distinction is a distinction in thought, not in the facts. It is a distinction which we introduce on reflection, not one given in actual experience. But though it is really quite untenable, it is also absolutely indispensable. I say it is indispensable. What would become of the whole theory of evolution, of the universality of the law of causation, if we tried to dispense with it? There is no need to try to dispense with it, if we remember that reflection is not the same thing as the experience reflected on. Remember that reflection is as it were the translation of the original experience, and it is easy to understand that the two things are different though related—

that what is one sentence in the original is broken up in the translation into two sentences connected by a conjunction, " and." Error only comes in when statements which are true of the translation are supposed to be true of the original—when it is supposed that the distinction which we make on reflection between cause *and* effect is a distinction which was given in actual experience.

Now, what I have said about the distinction between cause and effect will be found, I think, to apply equally to the distinction between the present and the not-present, *i.e.* the past and future—between the " now " and the " not now." What is included in the present moment? something or nothing? If nothing, then the present is a vertical line, a line without breadth—and that is an absolute abstraction, a pure nonentity. In that case the past and the future exist but not the present—and that is self-contradictory, for past and future, whatever they may be or have been, are now non-existent—the one no longer, the other not yet, existent. Still, the tendency is—especially on the part of Sensation Philosophers—to draw the line as fine as

84

possible. How fine then can the line which represents the present be drawn? Evidently the minimum below which we cannot go must contain at least a single sensation. The sensation which preceded the present sensation is past, that which is to follow is future. We look, then, as it were through a narrow slit, wide enough for just one sensation, which sensation must pass away before another can come. It is in their succession that past and present consist. But the past when past is not wholly past—it survives in memory, if not in fact. Now, one objection to this view is the discontinuity of experience which it necessarily assumes : we are looking through the narrow slit of the present moment and see the sensation before us, that sensation has to disappear, and the next to take its place. The break between the two is supposed either complete or not. If supposed complete it is untrue to fact : our life is not a series of detached moments. If the break is not complete, that is to say if there is no break, then we may compare the process to a diorama : as it moves before our eyes, part of the picture we have seen is gone, part of that which we are going to see is coming on. If then we are not aware of two successive sensations simultaneously, at any rate we are simultaneously aware of the latter half of the first and the earlier half of the second.

But, having got so far as this, having come to recognise that we may see together the end of the one and the beginning of the other, we may perhaps inquire why the range of our vision is supposed to be thus limited, and whether as a matter of fact it is thus restricted. It is supposed thus limited in the interest of a theory—a very natural theory—viz., that the distinction of past, present, and future is given to us in experience, not an interpretation of experience. Now, theory must fit facts : it is useless to mutilate facts to suit theory. The question therefore simply is whether we actually are aware only of one sensation at a time. We have seen that this is strictly untenable, even on its own pre-suppositions. Now let us set them aside and imagine we hear a burst of melody—whether from "a happy, happy bird" or a solemn organ's stately peal. Is it the case that we are aware of a succession of independent notes, and that having heard them one after another, we put them together and eventually on reflection get a distant idea of a tune ? Surely it is the other way about ! First the phrase is there, whole, complete, enrapturing. Subsequently, in cold blood we may dissect it into its component parts, its successive notes. But that is a subsequent proceeding. In other words the time-distinctions of past, present, and future are introduced by us, on reflection, into our experience : they are not essential to or necessarily a part of the experience. I say "the phrase is there," and I suggest that you were aware of the words together—that they formed a unity. You may now say there are four words: the-phrase-is-there ; and that " the " comes first, "phrase" second, "is" third, and "there" fourth. You may treat the separate words as though they were so many lantern-slides : you may put in the first, throw it on the screen ; take it out and put in the second, and so on, making each picture on the screen quite discontinuous with the rest. And then you may go on to argue that therefore the sentence as I originally uttered it was nothing but those four separate, discontinuous, successive words. Now to argue thus is to confuse the original, direct experience with the very different reflection upon it and dissection of it, which we may make but do not always or necessarily make. Yet the two things thus confused are really quite distinct and easily distinguishable. I may perhaps illustrate the distinction by calling your attention to a table : as it stands it is a table, but separate the four legs from the top, cast them apart ; and then, though you have five pieces of wood, you no longer have a table. Separate and apart from one another, the five pieces are not a table ; and, on the other hand, the table was not, or is not, merely five pieces of wood—it has a use, a purpose, a meaning which they have not. It has a being different from theirs. They and it are undeniably different. For one thing, it has a unity which they have not. So too, I proceed to argue, the sentence has a unity which its constituent words, when it is dissected into them, have not. That it can be dissected into them, we recognise on reflection and when our attention is called to the fact. But in the first instance, in our first immediate experience, it is the unity which is presented

to us, and of which we are aware. The
sentence as a whole is before us, and it
is all—at once—before us. " The phrase is
there." I submit that that sentence as a
whole, and the whole of. that sentence, was
grasped by you at once. The opposite, and,
as I am maintaining, the erroneous view, is
that each word successively alone was present
and presented, and that each word was past
and gone before you could be aware of the
next: through the slit of attention you could
only see one word at a time. What gives this
view its plausibility is that, when your attention has been called to the fact that it can be
dissected into four successive words, you can
easily repeat the sentence and note the words.
From this it is easy to infer that, when the
sentence was first uttered, you began by hearing
the words separately and then you constructed
the sentence out of them subsequently. But
the inference, though easy, is, I suggest, mistaken, and a reversal of the actual proceeding.
This, I believe, I can show quite easily. If
the inference is true of the sentence, it must
also be true of the words composing the
sentence. If you were conscious first of the

words, then of the sentence, and, last of all, of the meaning of the sentence, then you were also first of all conscious of the initial consonants, next of the long vowel, next of the sibilant, then of the whole word "phrase," and finally of its meaning. Now that, I think, would be plainly an error. Of the 800 millions or more of persons who people the earth, the vast majority cannot spell, and have no idea that a word can be broken up at all. Breaking a word up thus is a procedure of reflection; and to that reflection they never do, as a matter of fact, proceed. Breaking a sentence up is also a procedure of reflection, and a procedure to which countless millions of spoken sentences are never subjected. On reflection, indeed, we can say that the first half of the spoken sentence was past before the second half came to be uttered; or that, whilst the first half was being spoken, the second half was yet in the future. That is to say, we can on reflection introduce time-distinctions between the words of the sentence, or indeed between the syllables of one word. But in the first instance what we are aware of is the spoken phrase, or the phrase of melody as a whole—the phrase is before us as a whole : we dissect it, or may dissect it, afterwards. The distinction into past, present, and future is not something given to begin with : it is a distinction made by us—subsequently applied by us to what in direct experience was a timeless whole. Time is a form of analysis, a way in which we may analyse experience.

Perhaps you may feel that though the argument I have been setting forth has an appearance of plausibility when illustrated by reference to a simple phrase of words or notes, it can hardly be stretched far enough to embrace all time. If so, let me use a simple illustration. Imagine the capital letter V. Imagine the two strokes, however, prolonged to infinity. Imagine, further, that you are looking from the base of the letter. You can only see as far as your sight permits you, and those of us who are short-sighted cannot see very far. A line drawn across the V will mark the range of my vision, a more distant line will mark the range of vision of a better sighted person. Thus we may imagine a series of parallel lines drawn across the V ;

and each of these lines will differ in length from all the others. Now, the length of any one of these parallel lines gives, so to speak, the "time-grasp" of the mind. The "time-grasp" of a mortal man may be very restricted; but that of the Infinite and Eternal Mind of God will comprehend infinity. To His comprehension infinity will be what a short sentence is to our minds—as comprehensible as the words I AM. But time-distinctions, though they may be applied to infinity and the Infinite, are evidently inadequate to cope with it; and that inadequacy is a further confirmation of the view that time-distinctions are not something given to begin with, but are distinctions made by us. We may seek to interpret Eternity by means of the time-distinctions we employ, but we cannot get Eternity into them, or stretch Time wide enough to include Eternity. Useful as time-distinctions are, they are also a method of interpretation used by us—they are not the reality which we interpret by means of them. It is not in them, nor are they in it. We translate Eternity into terms of Time; and so familiar are we with those terms, that I feel as though it were paradoxical to state the simple truth, viz., that Eternity is real and Time is not.

The paradoxical appearance of the statement that time-distinctions are only a method of interpretation used by us becomes very pronounced when we set our faces towards the future. Surely the difference between present and future is a real difference, a difference in things, and not merely a difference of interpretation? We may know the past and see the present : we neither see nor know the future. We have present sensations, memory of past sensations and expectation of future sensations—expectation but not knowledge. Well! then, can we be said to have knowledge of the past? If it is admitted that we can, then we can have knowledge of what is not present. Now, past and future alike are not present ; but the suggestion made is that of the past we have memory and therefore knowledge ; while of the future we have expectation and therefore not knowledge. How far then is memory the same thing as knowledge? Some people remember things which I will venture to say—they not being present—never took place. On the other hand, when we base ourselves on the Nautical Almanac, is our expectation of the next eclipse inferior as knowledge to our memory of childhood's days? Or is the proposition, "all men are mortal" of less certitude when applied to the future than to the past? If it is equally certain, then we have some knowledge of the future. Though the moment of death is not known, even in the case of condemned criminals, quite so accurately as that of the eclipse of the sun, there is as much certainty about it as about most of our memories of the past. And generally speaking, we rely with justifiable confidence on lunch being served to-day at much the same time as yesterday. There is probably as much correctness in your anticipation of this afternoon's proceedings as in your re-collection of the events of yesterday afternoon. Or, to take a wider view, our earthly life lies between birth and death, the one event in the past and the other in the future: the future event is as certain as the past event, and the past is not more matter of knowledge than the future. If then we liken the brief span of life to the phrase that is spoken or sung, we may comprehend that the time - distinctions with which we may and do, on reflection, break up the unity, whether of the sentence or of life, are introduced on reflection, and are not there in immediate experience. The difference between present and future which seems so undeniable when once our attention is directed to it, is a difference which does not exist until our attention is directed to it. When our attention has been called to it, we can repeat the sentence which as first uttered was a unity, a whole, in which there were no time-distinctions and no consciousness of any such distinctions. We may stop in the middle of the sentence—we may say "the phrase" and pause. Then the pause represents the present, the words we have uttered are past, and the words we are going to utter—" is there "—are future. But until and unless we thus introduce time-distinctions, the sentence is timeless. Until and unless we introduce time-distinctions, consciousness is timeless. Time is a way in which we may interpret consciousness. But the fact that Time is in the interpretation or is the form of the interpretation is no proof that it is in the original.

The difficulty we undoubtedly feel when we try to realise the timelessness of experience and consciousness is, I think, due to the fact that in order to interpret consciousness into terms of time we have to assume that we are here and now. When, but not until, we make that assumption, we can distinguish the "now" from the "not now." We can divide the "not now" into the "no longer," and the "not yet." Then, if it is suggested to us that there are difficulties inherent in this process of distinction, we are apt to understand simply that the "not now" is unreal. We try therefore to abolish the "not now" from thought ; and we find ourselves imagining Eternity as a "now" prolonged to infinity both ways. The result is that, after all our struggles of this kind, we do not get rid of time-distinctions. Whether the "now" is conceived as a narrow slit through which we see the present sensation—even though the slit be imagined so narrow that it is length without breadth—or whether the "now" be expanded to such a width that it is infinite, in either case so long as we start from the "now" there must be a "not now" outside it. And so long as that is the case we evidently have not reached the notion of Eternity : we are still within the limits of time-distinctions. Evidently, therefore, the conception of Eternity is not to be reached this way—it cannot be reached, that is to say, by insisting on the "now" and denying the "not now." *Both*, alike and equally, must be denied and set aside. In a word, Eternity is not a "now" infinitely extended : it is timelessness.

If it should be felt that this idea is extravagant, I would point out that at any rate it is not self-contradictory. And I should like to go on to say that time-distinctions are plainly self-contradictory ; and that the only way of getting over the self-contradictions is to recognise that the distinctions are themselves unreal. Then, if they are unreal, we may perhaps understand that the self-contradictions are due simply and necessarily to the original error of mistaking an unreality for a reality. I say, then, that time-distinctions are unreal, that they are self-contradictory. This suggestion I have in effect already made when

RELIGION IN EVOLUTION

I pointed out the difficulties attendant on attempts to define the "now" or to indicate its extent. What extent do you assign to the "now" to "the present time"? You may perhaps vaguely mean the present hour of lecture. You may, of course, mean the present moment—which, of course, is no longer present—it was over before I could speak of it. Indeed, when you come to look at it, it was but the line dividing the past from the future—and a line, as you know, is length without breadth—that is to say, it had no breadth, that is no duration—it had, in fact, no existence. Length without breadth is purely imaginary. The present, in that sense of "the present moment," is not a real thing. Now, that conclusion is exactly in harmony with the view that I am engaged in maintaining, viz., the unreality of the "now" and the "not now." Past and future time, for those who believe in time, must essentially be the same as present time. The past is made up of moments that have gone by ; the future of moments yet to come —and those moments, like the present moment, are imaginary. If you exist in the present, if your past existence is over and now non-existent, if your future existence is not yet and may perhaps never be, then you exist in the present, that is to say you have to squeeze yourself up within the limits of a line which has no breadth, and is, in fact, purely imaginary. Personally, I don't feel that I have room to breathe when laced up so tightly. Shall we then relax the limits of the "now," and leave ourselves room to breathe and turn in? We have seen that we can speak intelligibly of the present hour. We can speak with equal reason of the present day, the present century, the present age, the present epoch, the present dispensation. When we do speak and think thus, "the present" seems to have widened itself out as unreasonably as the "now" narrowed itself a moment ago when it shrank to a line's breadth, *i.e.* to a breadth which is purely imaginary, as we have seen. Perhaps you will say that, of course, when we speak of the present century we don't mean that it is present. Of course, too, when we speak of the present day or hour we don't mean that it is present. Or, if we come to that, the present minute with its sixty seconds

is not "present." On the whole, then, it
would seem that "the present century" or
"the present hour" is not a real thing, any
more than "the present moment" is. There,
too, once more I quite agree. "The present,"
so far from being the one actual reality in
which alone we live and move and have our
being is a self-contradictory idea. There is no
fixed or reasonable line to be drawn between
the "now" and the "not now." We move
the line about to suit our own purposes. The
"now" and the "not now," each of them in
turn, threatens like a dragon to swallow up the
other, and seems to have done it—but never
does do it, for these are relative terms, and
neither can enjoy its shadowy existence without the other. What we have to do is to recognise that these monsters—and how horrible
at times is the difference between the "now"
and the "not now"!—are not realities, but
shadows which the blade of reason may strike,
and not in vain. And if it does strike, it strikes
them both; and when they vanish Eternity
arises—not an everlasting "now," but pure
timelessness.

Let us now turn from Time to Space, and

let us consider whether space is one of the realities of which we have experience, or one of the ways in which we may interpret the experience we have. To begin with, it is clear that some of the arguments we have used—whatever their value—are as applicable to space as to time. As the moments of time are relative to one another, so are the points of space. As the "now" is relative to the "then" and is intelligible only in its relativity, so the "here" is relative and intelligible only as relative to the "there." If there were no "here," there could be no "there," just as there can be no "then" or "not now," if there is no "now." Again, there is no peace between the "here" and the "there," any more than there is between the "now" and the "not now." In turn the "here" seems to swallow up the "there," and then to shrink into nothingness itself. Let us try to define "here," or at any rate to form some notion of it. It would be quite correct to say that you and I are "here," meaning in this room. It would be equally correct to say that we are "here," meaning in this town, or—wider still

RELIGION IN EVOLUTION

—meaning in this England of ours ; or—wider still—in Europe ; or in this world ; or—widest of all—in this universe. " Here " is in fact a word which, like a stone thrown into a pond, sends out widening circles, which go on spreading until they embrace everything : what at one moment was outside the circle and " not here," at the next is " here "—and so on, until everything is " here." The " there " or the " not here " seems entirely abolished and swallowed up. It seems, but only seems, abolished : the seeming is mere appearance. Consider ! by " here " we may mean " in this room " ; but equally well it may mean " at this table," or " on this paper." And there it may mean " this word," " this letter," " this letter i," or " the dot upon the i." And when we have got to this dot, we may remember that the dot is a point, and that a point is position without magnitude. " Position without magnitude ! " the thing is an unreality. It is not a thing of which we have experience. And space is an arrangement of points, that is, of unrealities. Sum up all the unrealities, and though they be infinite in number, the sum is as unreal as the items which make it up. The " here," having swallowed up the "there," now shrinks and shrivels up into nothingness.

Space then alternates between being a mere blank, in which all things may be, and being a mere point of nothingness, from which all things are excluded. And so long as we imagine space to be real, we are compelled to vacillate between these two extremes, and perpetually to abandon each in favour of the other. Its infinity is illogical, incomplete, unsatisfying. It is illogical, because it is impossible to have one of two relative terms without the other—impossible to conceive of a "here" to which there is no "there." It is incomplete, because, extend it as you may, you cannot extend it so far that there is no "beyond." It is unsatisfactory because you *cannot* extend it so far, and yet feel that you must. The escape from this unsatisfying want of logic lies in recognising that space is not a reality of which we have experience, but a way of interpreting the experience we have. It is a way of interpreting the unity of things, the unity of which we are aware and of which we form part. But, like all interpretations, it consists in substituting for the original a version which, however good, however useful, is not the original. It is reflection, and the outcome of reflection upon the original. But it is not the experience which it dissects, or rather of which it is the dissection. The dissected creature is not the living creature : the unity of life has gone. Space is a form of thought by means of which we break up the unity of experience, and into which we distribute things when under the influence of reflection their unity has evaporated. The difficulty, or rather the impossibility, of piecing together the " here " and the " there," of defining their edges, so to speak, in order that we may set the broken bone, is itself the proof that the unity of life, of reality, is not in them. By means of the conception of space we hold, or imagine we hold, things apart, so that we may contemplate and examine them separately. But the moment we have ascribed to them this separate, individual existence, we find that, when we want to put them back into relation with other things, those other things also are in space. The result is that we do not get back to the original non-spatial unity of experience : we find ourselves in a spatial world, containing separate things, effectually isolated from one another by their spatial relations. The only unity those isolated things possess is such unity as they can attain in virtue of the fact that they are all in space. And how unreal is any unity which professes to unite the "here" and the "not here," we have already seen : it is as unsatisfactory and impossible as the attempt logically to distinguish them.

Thus far we have been considering arguments which apply equally to space and time, and calling attention to points which space and time have in common. But it is equally true that there are points of difference, and it is equally necessary to consider them. The first point I would call your attention to is the fact that space bears a special reference to my individual, personal experience, which time does not. This present moment which is "now" for me is also "now" for you, and for all in this country or this world. But "here," the space which I occupy, is not "here" for you or anybody else—it is "there" for everyone but me : two things cannot simultaneously occupy the same point of space, but an infinite number of events or experiences may and do take place in the same moment of time. If we may take "now" as the central point of time, from which we look back upon the past and forward into the future, then we are all of us alike and equally at that point, in the "now." But with the central point of space the case is different. Space has its centre everywhere and its circumference no-where. It has its centre everywhere, for wherever there is a centre of consciousness there is a centre of space, that is to say, a "here." Its centres are infinite in number and never coincide. But its circumference is no-where. That seems to mean that it has indeed a circumference, but that that circumference is not in space, or perhaps, rather, we should say that what lies outside the circumference is not space. But in what intelligible sense can we speak of space as being in the non-spatial? If it is implied that the non-spatial does not exist, the position is open to all the difficulties and self-contradictions involved in the mean-ingless phrase "infinite space." If it is implied that the non-spatial does exist and is a reality, then the spatial does not exist, or at any rate is a non-reality. In truth we have travelled, as all who talk about space must perpetually travel, from one extreme to the other. We narrowed down the "here" until it was position without magnitude; and now we have widened space out, until, with its centre everywhere and its circumference nowhere, it is simply magnitude with no position.

I have said just now that wherever there is a centre of consciousness there is a centre of space, and that space bears a special reference to my individual, personal experience. And the consequence of starting from these assumptions, viz., that there are "centres," that is to say local centres, of consciousness, and that individual experience is the bed-rock, or the seed-plot of experience generally, is that we are eventually landed in an "antinomy of thought," that is to say, we find ourselves compelled to say that space must have, yet cannot have, bounds. But though we may say that in words, we cannot profess that the words have any meaning. The fact which comes out plainly is that the very idea of space is self-contradictory and unintelligible. Under those circumstances, it is I think not unreasonable that we should go back to the premises from which we started and examine them again, and see whether they are really sound.

When we speak of "centres of consciousness" and mean thereby local centres, it is obvious that we are assuming that space really exists. And that is an assumption which we cannot properly make when we are engaged in enquiring whether space does exist. Further, we spoke of "individual, personal experience" as though each individual's experience were a separate world by itself, which could exist quite independently of all other persons and beings, and would go on existing if all other spirits, even God Himself, were not. Now that is a view which is held, it is the philosophical system of "solipsism." It is a view, however, which I shall set aside, on the ground that we can and do share each other's experience, or rather that there is a common experience in which we all share. Now, this community of experience, or communion in experience, cannot and does not override the conviction each one of us has of his own personal individuality. And having granted this, having insisted on this, I wish to go on to ask whether it follows that our individuality, our centre of consciousness, must be a local centre—a centre in space. If it does, then, as I have pointed out, it follows that there are many centres, and that the circumferences are not real. And if the circumference of a circle is not real, neither is the centre : a circle which has no circumference can have no centre. It simply is not a circle at all, it is a self-contradiction. Well! but does it follow from the conception of personal individuality that such a centre of consciousness must be in space? If it is in space it must occupy space, and the experiences within it must also occupy space. If you are your body and nothing else and nothing more, then you do occupy space, and "shooting pains" may travel through your body, traversing and occupying space. But if that is so, then the thoughts that go on in your head, or the pain that rages in your tooth, also occupy space. Now, you localise the tooth-ache in that particular stump ; as a matter of fact it is *not*. Persons who have had a foot or a leg amputated still, when the weather changes, locate the pain they feel in the corn which is no longer there, though they could swear it was. The pain is there, but it is not in the spot in which you locate it. Perhaps, then, though it is not in that particular spot, it is in some other point of space. Well! if it is in space it must occupy space. And how much space does a raging toothache occupy? Miles! Neither lineal measure, no, nor cubic measure will express it. Great thoughts doubtless are the prerogative of great minds; but does any one seriously imagine that thoughts measure two feet by three, or that great minds can be estimated by cubic contents? It is obvious that pain and joy, thought and resolution are non-spatial. They do not occupy space, and thus cannot be in space. They are not in space, they are in you and me. Then, neither are you or I in space. If the pain which I have does not occupy space, neither can I who have the pain occupy space. Space is not something in which I am, but a way in which, or a language into which I interpret my direct experience. To speak of myself as a centre of consciousness, and to ask whether that centre is in space, is to use a metaphor and to ask whether it is actual fact and truth. If it were, it would not be a metaphor. But it is a metaphor, and a misleading metaphor. To admit that thoughts have breadth or width or depth only in a metaphorical sense, that, in fact, they do not occupy space, and yet to imagine that they are inside your head somewhere is a logical impossibility. It is really a repetition of the fruitless journey which we have travelled already. We were told to conceive space as a circle, with its centre everywhere and its circumference nowhere; and then, outside the circle, we had to conceive the non-spatial. Now, the fallacious argument exhorts us to seek the non-spatial not outside the circumference, but inside the centre. This time, however, the non-spatial includes thought, will, emotion—all experience. What remains may be shot down in space. Matter may be deposited there.

The reason why I have devoted a lecture to the subject of time and space may now be stated, if indeed it is necessary to put it into explicit words. It is obvious that Evolution is a process conceived to take place in space and time; and therefore the Evolution of Religion is a process conceived to take place in space and time. But if space and time are not ultimate realities but ways in which, or forms according to which, we interpret experience, then space and time are part of the translation, or the form which the translation takes; but they are not in the original. Evolution is not the ultimate fact with which we have to deal. The theory of Evolution is a way of re-arranging—in thought—our experience of fact, or perhaps I should rather say our experience which is fact. That re-arrangement consists in a re-distribution of experience, in parcelling it out as occupying this portion of space and that period of time. So far common-sense detects nothing in the arrangement with which it disagrees : science is and boasts that it is nothing but common-sense, clarified, it may be, and consistently applied. Even the principle by which science accounts for or explains the changes which it conceives to occur in time and space is a principle which it owes to common-sense, and which is therefore approved by common-sense—it is the principle of cause and effect. But the lengths to which science carries out this principle in the interests of the theory of Evolution are such that eventually common-sense must revolt from them ; for eventually the theory of Evolution seeks to exhibit everything that is and occurs as subject to a law of Universal Causation. If everything that is done or that happens is the inevitable effect of a pre-determined cause, then everything that is done by you or me is the necessary outcome of the causes at work; and our freedom, the freedom of the will is gone. But there common-sense revolts. Science and the theory of Evolution act very well so long as we exclude the existence and activity of free moral agents from our view; and only so long will they act satisfactorily. In other words, science and the theory of Evolution are abstract: they are abstractions from experience, they are only partial views of experience, and they are views which can only be got by closing our eyes to the existence of free agents. It is obvious therefore that science and the theory of Evolution afford only a partial explanation of the Universe. They do not aim, and avowedly do not aim, at more. But in that case they can never give an explanation of the whole. If the interpretation which they aim at putting on the facts with which they deal were to become absolutely

H   113

exhaustive, it would still leave unaccounted for the existence and the action of free moral beings. The interpretation might be a theory of Evolution, complete at all points : it would not be a Philosophy. And so too, however complete a theory we may get of the Evolution of Religion, it cannot, so far as it is Evolution, be Philosophy.

I have said that the theory of Evolution gives only a partial explanation of the Universe ; and I mean to imply that a partial explanation is something very different from the explanation of a part. If instead of being very different they were the same thing, then science might rest safe and satisfied : her own garden-plot would be marked off by itself, and she could cultivate it without danger of being interfered with. Science and the theory of Evolution are abstractions from experience : they are abstractions and not realities ; and they are abstracted from the experience of free moral agents. That is to say, science deliberately and rightly ignores the fact that, throughout, it is dealing with the experience of free moral agents ; her object is a partial explanation of that experience, an explanation of it when it is viewed as something with which free moral agency has nothing to do. Error comes in only when it is alleged or understood that there can be something in the experience of free moral agents which is entirely independent of and aloof from that free moral experience. The error becomes glaring when it is supposed that the ultimate explanation of that experience proves that neither freedom nor morality is in it. There, as I have said, common-sense revolts : morality and the freedom of the will it will not part with. Then it must part with the universality of causation. And if the law of cause and effect is not universal —if, for instance, the will of a moral agent is not subject to it—then the law of causation is but a mode of interpretation : it is essentially not the explanation of a part of our experience but a partial explanation of it. It is a law which requires time and space to act in ; and like time and space it is not a reality or an experience, but a way of interpreting reality or translating experience. The fact that time and space and the law of causation afford but a partial explanation of reality, and a

limited explanation, is shown by the anti-
nomies of thought implied by them. To seek
the first of causes, the end of time or the limits
of space is an endeavour which we can only
renounce with satisfaction when we recognise
that time, space and cause are not realities
—that they only occur in the translation and
are not to be found in the original.

IV

EVOLUTION, we have seen, assumes the reality of time and space, and the validity of the law of causation. It sets aside the freedom of the will and ignores the possibility of the existence of God. It may or may not be right in making these assumptions; or rather it would be more correct to say that it is both right and wrong in making them. It has a right to make them, inasmuch as science has found by experience that the best way of attacking complex problems is to simplify them artificially, that is to say, to concentrate attention on some one aspect of them, and to deal with that in abstraction from the rest. When we have learnt how certain factors would behave —what results they would produce—if they were the only factors, we are the better able to judge of their action and effect when they are complicated with other factors. The possibility of error arises when it becomes doubted whether there are actually any other factors to take into account. When such doubt arises, the question debated is whether the conclusions of science are abstract—and as abstract require correction—or whether they are not. That is the question which is always implied, though not always recognised, in debates as to the relation of science to religion. Historically indeed there is no doubt that men of science began by simply claiming provisional freedom for science : they claimed that they should be allowed to cultivate their plot in the garden of knowledge without liability to irruption and invasion from theology. And not only have they claimed this right, they have established it : they have repelled the onslaughts which have been made upon their domain, and have repelled them so triumphantly that they have peace upon their borders —or might have peace if they chose to stand purely upon the defensive. As a matter of fact, however, they do not always abstain from reprisals : attacks upon religion in the name of science were in the latter part of the past century as frequent as ever attacks upon science in the name of religion had been

RELIGION IN EVOLUTION

— and much more telling. Indeed we might go so far as to say that science has not only invaded her neighbour's territory but claims to have annexed it : the Science of Religion is but one province in the empire of science. The claim, however, is not admitted by the inhabitants, nor do they render allegiance to the invading power. The title-deeds of the claimant, so to speak, are called in question. The claim would, or might be valid, if the assumptions on which it is based were proved to be true. Those assumptions —with regard to time and space, the law of universal causation, the freedom of the will, and the existence of God—are assumptions the validity of which requires to be proved, and can be tested only by Philosophy. Now it is, I believe, not going too far to say that men of science are beginning to recognise that no proof of these assumptions can be given. All these assumptions are of the nature of hypotheses ; and any hypothesis is now recognised by science provided that it is capable of explaining the facts which require explanation. And only so long as it does explain them is it thus recognised :

the moment it fails, or a more comprehensive hypothesis emerges, that moment the old one is thrown on to the scrap-heap of science. Thus science is coming to be consciously hypothetical, and to be aware that she is purely hypothetical. All that she requires of her hypotheses is that they should account for the facts: verification, in the sense in which that word is defined in Mill's " Logic," science does not now profess to attain or to even aim at. The point, however, to which I wish to call attention is that the facts for which science undertakes to account are facts of human experience; that those facts are — in quite a legitimate manner — artificially simplified by science; that they are simplified because they are taken in abstraction from the experience of which they are part ; and that in particular they are taken in abstraction from the freedom of the will and from the existence of God.

Now, there are those who hold that the will has no freedom and that the world has no God; and for them, therefore, it might appear that science is no abstraction from experience, and requires no correction—that it is the truth, and the final truth, of experience. A moment's reflection, however, should suffice to show that this is not quite the case. Science and the scientific theory of Evolution are not built upon a denial of the existence of God or of free - will : they do not require us to begin by denying either; they simply require us to leave aside both for the time being. The object of science is not to enquire whether either is a reality, but to build up the theory of Evolution in such a way that it cannot be affected by any views we hold or conclusion we may come to as to the existence of free-will or God. The assumption seems to be that knowledge or experience is divided, as it were, into water-tight compartments, none of which can be affected by anything that goes on in any other. On this assumption it is imagined, by those who make it, that liberty is secured both for religion and for science : each line of thought may be produced ever so far both ways, and the two lines will never meet or clash. Now, this assumption is obviously fatal to the belief that science is the truth, the whole truth, and the final truth of experi-ence : it is fatal because it sets up religion

as being actually or possibly as true and real
as science itself. And it appeals very strongly
to the sense of justice in men of science : they
have claimed for themselves that science
should be free to run its course without
interference from religion, and it is only
just that religion should be allowed the
same liberty and should enjoy the same
freedom from interference on the part of
science. It is, however, tolerably clear that,
just as this arrangement appears to be, and
indeed may be, it is an arrangement which
has no element of permanence in it. It sets
up religion and science as two masters of
equal authority, and it offers no reason whatever for believing that it is impossible for them
to come into collision : it simply assumes that
they can never meet or clash. Either that
assumption may or may not be discussed : if
it may not, it is pure dogmatism ; if it may,
then it is for Philosophy to discuss the relation of science to religion. In the same way
the assumption that science is no abstraction
from experience, and therefore requires no
correction, is either pure dogmatism or else
it is a proposition which is open to discus-

sion—and the discussion is Philosophy. The question whether science is or is not an abstraction from experience is simply the question whether the freedom of the will and the existence of God are facts of which we have experience or not.

Perhaps I may put this point in another way. Science and the theory of Evolution are built upon the understanding that science must go on its way quite unhampered by the question whether there is or is not a God. As far as science and evolution are concerned, that question is not raised : it is assumed that we do not know, and for the purposes of science do not require to know. And so long as we adhere to that assumption, the position of science remains unmoved. But the moment that we know, or think we know, the position of science becomes altered. The position was that we did not know, the fact is that we do know. And the position of science necessarily becomes altered by that fact. Whether we believe that there is or is not a God, the position of science is bound to change. In the one case, science ceases to be an abstraction from experience and becomes the whole and sole truth of experience. In the other case, it is seen that science is not the whole truth of experience, but an abstraction, and an abstraction which will require correction before it can take its proper place in our experience of the real. Science does not know whether the will is free or God exists. If we know both facts, then our view of science and of the theory of Evolution will be very different from the view of the man who knows neither.

I have spoken of the freedom of the will and the existence of God as facts of which on one view we have, and on the other we have not, experience or knowledge. And perhaps it may appear to you, as it has done to many, that our knowledge of the existence of God is a matter of inference. If so, I feel bound to point out the lesson which is taught upon this point by the history of Philosophy. That lesson is that all the many attempts to infer His existence have failed thus far. That fact should by itself, I think, suffice to give us pause, if we are inclined to renew the attempt to draw so great an inference. It should set us enquiring whether there is anything in the nature of the attempt itself which necessarily forbids the attempt from being successful. I suggest that there is something of the kind. If the existence of God is a matter of inference from our experience, then it is not a fact given in our experience. That in itself is an assumption—and an assumption against which some of us at any rate will rebel. But though we feel inclined to revolt against this assumption, let us recognise that it is made with a good object. It is made for the purpose of showing that we can, at any rate, infer His existence. Very good! what is the value of the inference? In other words is it a hypothesis of the same nature as the hypotheses of science, which are avowedly incapable of verification and are announced to be purely working hypotheses which will be cast aside as soon as they have served their turn and will go to augment the scrap-heap of science? If it is an unverifiable hypothesis of that kind, it has no meaning for us—we have no use for it. It might be scientific: it is not religious. Perhaps, how-ever, a student of the Evolution of Religion will say that after all it is precisely this that meets him in the history of religion : what he finds is that all sorts of hypotheses are, and have been, held and then rejected. The worship of animals, of the heavenly bodies, of the earth, of the gods of Greece — what were all these worships but hypotheses as to the being of God? And if they have gone down, what is that but a proof that they were unverifiable hypotheses? Religion, no less than science, has its scrap-heap, and is continually augmenting it. Now this is precisely the sort of difficulty and danger to which a student of the Science of Religion is exposed. It is almost inevitable that he will draw the conclusion that as it is with the history of science, so it is with the history of religion : in both cases we have to do with hypotheses, of which some have been cast aside, and the others will as certainly be sent to join them. In both cases we have to do with hypotheses or fancies, but not with fact. Now, it is obvious that in the worship of the sun, moon, stars, animals, etc., we have to do with fancies or hypotheses. But we also have to do

with something more. It is true that the
peoples of the earth have sought and do
seek their God in stocks and stones. But
it is also true that they cease to identify
Him with the tree or the animal in which
once they sought Him. The hypothesis
that He is to be found in the animal or at the
altar or in a house made with hands is abandoned in some cases and may be abandoned
in all. But though these hypotheses are or
tend to be abandoned one after another, they
are simply hypotheses as to the way or
shapes in which He manifests Himself. Such
hypotheses are very different from belief in
Him. Indeed the very fact that they change
is proof that He abides. It is because the
belief in Him is there all the time that the
hypotheses as to His shape or place can
change : unless the belief were there, they
would not change, they would simply cease.
The testimony of the Science of Religion is
that the belief is simply ineradicable. It is
quite distinct from the hypotheses as to His
place, shape, or mode of manifestation ; and
the question before us is as to the nature and
validity of the belief. Is it a matter of in-

ference? If it is, then it is all-important to us to know whether the inference is verifiable or not. If it is not verifiable, then an argument which represents the existence of God as something which is not matter of experience, but of inference, and proceeds to show that the inference is an unverifiable hypothesis, is an argument whose nature casts a doubt on the assumption that the existence of God is not a matter of experience. On the other hand, if the inference is verifiable, of what nature is the verification of which it is susceptible? The verification must be something of which we have experience. If astronomy infers that an eclipse will be visible at a certain time and place, the verification of the inference is to be found in experience. So too, if the existence of God is an inference, it is only by experience that the inference can be verified. But when it is verified thus, it is an inference no longer—it is a matter of personal experience. It is no longer a proposition dependent for its truth on some other proposition. It is not a hypothetical or conditional conclusion, true if the protasis or the premises be true; if otherwise, false. It is not a conclusion or an inference at all for those who have had the experience: it is a fact—and a fact of experience. That is why I suggest that all the attempts that have been made to exhibit the existence of God as an inference have failed and must fail. It is believed, when it is believed, as a fact, not as an inference. That is the reason why all the attempts to exhibit it as an inference which we must draw and cannot help drawing if we reason logically, have failed as all such attempts must fail. They fail in a sense discreditably because they begin by making an assumption the truth of which they are bound eventually to deny: they begin by assuming that the existence of God is not a fact of experience, and they end with the conclusion that it is. They begin with the promise that they will show His existence to be a matter of logical inference, and in the end it is found to be a question of personal experience, and not of inference at all. If all the attempts which have been made to exhibit the existence of God as an inference from certain facts have failed, it is because there are no premises vast enough or adequate to bear the weight of such an inference. The existence of God is the premise from which all things must be explained : they are not the logical condition on which His existence is dependent.

What then is the position in which we find ourselves if we admit that it is impossible to infer the existence of God? Obviously we have to give up the idea—attractive as it appears or has appeared in all probability to all of us at some time—we must give up the idea that it is possible to construct an argument which shall by mere force of logic make the existence of God an inference which a man, even against his will, must draw. Next, logical constraint of this kind, as it is impossible, so also it is superfluous : they do not infer Him who believe in His love and His goodness—they know His loving-kindness as a matter of their direct personal experience. On the other hand, to those who do not believe, this belief of ours appears to be an assumption. That we must freely admit ; and in admitting it we gain the right to say, not only with complete confidence but with perfect justice, that the other first principle, viz., that there is no God is also an assumption—and is an assumption which is at variance with the facts of our experience.

In thus declining to make the existence of God a matter of inference from experience, in insisting that it is and for those who believe can only be a fact of experience, we base ourselves on experience, and it becomes therefore necessary for us to see that experience is not defined in such a way as to beg the question. Indeed it is equally necessary for both parties to the dispute : if it is settled *à priori* that experience is only experience of tangible, visible things, then the existence of God cannot be a matter of experience ; or if it is settled *à priori* that experience is only experience of finite spirits, then we cannot have experience of God—the question is begged, for the whole question is whether we do have such experience. And it is our duty to protest against such a *petitio principii.* On the other hand, we have no right to begin by assuming that every finite spirit has conscious experience of God : we are bound to accept the evidence of those who testify that they have no such experience. And we may accept it without in the least binding ourselves to hold that they never will have, still less that they never can have such experience.

It is necessary to insist upon caution in this respect, because these are arguments which approach the Being of God from the point of view that that Being is a principle of unity. In the hands of some philosophers those arguments proceed to the conclusion that the principle of unity which binds free, finite spirits together is itself a Personality—the Personality of God. In the hands of other philosophers, however, this further step is not taken. By such a philosopher it is or may be argued that finite spirits may be united by a principle which, since it unites them, is a principle of unity but which is in no sense a person or a personality. The members of a college or a football club have, as a college or a club, a unity : they have common purposes, a common principle. It is that principle which gives them their unity : it is a principle of unity. The principle dwells in them, in each and all of them. Its reality is efficacious and undoubted. The spirit of the college or the club dwells in each member and works and manifests itself in each and all. But though the principle of unity which inspires and penetrates them all is a reality, no one I suppose would maintain that the principle in such cases is a person or a personality. This mode of argument may then be applied, has been applied, to that principle of unity which binds together all free, finite spirits. That there is such a principle, thus uniting all the free, finite spirits of the whole universe, is conceded or assumed. But the principle of unity in their case is taken to be the same as the principle of unity which binds together the members of a college or a club. In the latter case the principle is not conceived to be personal: in the former case therefore the principle of unity which binds together all thinking beings is not a personal God but an impersonal Absolute. This argument is set forth in Dr M'Taggart's "Hegelian Cosmology," and he sums it up by saying (p. 94): "I think therefore that it will be best to depart from Hegel's own usage, and to express our result by saying that the Absolute is not God, and, in consequence, that there is no God."

If Dr M'Taggart's argument is logically

correct, it will be a fresh instance of the position I am maintaining, viz., that all attempts to exhibit the existence of God as an influence from experience are foredoomed to failure ; and I do not feel concerned with it further from that point of view. The reason why I have alluded to it is that it shows how necessary it is to insist on the existence of God as being actually or possibly a fact of experience and not an inference from experience. If we start from the experience of free, finite spirits, and seek to discover what there is in it or may be inferred from it, we must either take or not take the knowledge of God as a fact of experience. If we do so take it, then no argument is necessary to prove it or capable of proving it. If we do not so take it, then we may go on whithersoever the argument carries us—and wherever we are* wafted we end with an inference. The inference may be that the principle of unity which binds together all thinking things is a personal God, or that it is an impersonal Absolute, and that " in consequence there is no God." The latter negative inference will have no hold over those whose belief is based not on inference but on experience. And the positive inference will not be satisfactory even to those who seek rather to infer God than to know Him.

I have insisted now so much and so often on the fact that the existence of God is not an inference—not a fact capable of being inferred —that I feel I have earned the right to say that in exactly the same way there is no possibility of logically demonstrating the non-existence of God. I think the idea generally is that no one can prove His existence, and that, in the absence of proof, the inference that He exists is illogical and only tenable by unreasonable minds. It seems a perfectly fair and indeed proper position to take up, to say that one is prepared to believe anything that can be proved, and that one cannot be expected to believe things that can't be proved. To ask for proof, to demand the premises from which a conclusion is drawn, to require to satisfy one's self that the rules of the syllogism or of induction have been complied with—all that is reasonable and praiseworthy. The question is, How far can the process be carried on, or carried back? If only propositions which can be inferred are properly to be believed, then every inferred proposition is a conclusion from premises which in their turn are inferred from previous propositions—and so on. Yes! so on. But how far? If we go on thus *ad infinitum*, we have a chain of arguments hanging down as it were from the sky. The bottom end we have indeed in our hands : but the other end—if there is another end—is out of sight, and we don't know whether it is firmly fixed up or will come down with a run on our heads. Evidently we cannot believe that the regress of inference is infinite : we must assume that there are ultimate major premises of all demonstration. But if they are really ultimate, and are also worthy of belief, then the whole of our inferred knowledge depends upon propositions which are not inferred. It would seem, therefore, that a proposition may be worthy of belief even if it is not an inference, and that inferred propositions owe their validity to the fact that they are inferences eventually from propositions which are not inferred. Now, propositions which are not inferred are sometimes spoken of as facts or facts of experience. And with one reservation, or rather explanation, we may accept the statement. The reservation is that all so-called facts are propositions. It is sometimes assumed that a fact is something corresponding to one of the terms which make up a proposition; and on that assumption it is supposed that we can make a bridge over from an outside world consisting of loose, unconnected or disconnected "facts" to the world of experience in which our thought, and we as thinking creatures, live and move and have our being. But if there be such an outside world of such loose, disconnected facts, we at any rate have no knowledge of it, nor does it come within our experience. If we think, we think propositions. A term is either understood by us to be part of a proposition expressed, understood, or implied, or else it is a meaningless sound. If it is more than a meaningless sound, it is a sound with a meaning, and the meaning is a proposition.

I will take it, therefore, that in experience we have to do, and can have to do, only with propositions, and that of propositions some are inferred from other propositions and some are not. There is, therefore, nothing irrational in

that there are other thinking beings besides ourselves, and we have assumed that those beings have a common experience—that there is not only a plurality of spirits but a communion of spirits. That these assumptions are true is questioned by the philosophy of Solipsism, which asserts that I exist and that there is no proof that other beings exist, still less that they and I have communion. Once more then, if Solipsism is right, we can proceed no further; and if we decide that we must, or at any rate that we do, and that we will, believe in a community of spirits having a common experience, we can only do so as a matter of faith. And that there is truth, one and indivisible and valid for all spirits, is as I have said also matter of faith. From this point of view, viz., that there is such a thing as truth—a point of view which after all is the ordinary point of view—it is inevitable that it must either be or not be true, and cannot both be and not be true, that there is a personal God. Further, every man who accepts one of these alternatives on the strength of his own experience must regard the other as a mere assumption. In other words, he assumes the alternative which he accepts to be that which is true for all thinking beings : he is asserting not only that it is true for him, true in his experience, but that it is the truth of common experience, of that experience which is common to, and shared in, by all finite spirits. So long, however, as others make the opposite assumption, as they have a perfect right to do, so long he must recognise that his belief is for others what theirs is for him, viz., an assumption.

It comes to this, then, that it is impossible to demonstrate either the existence or the non-existence of God as a logical inference from any premises that we have or can imagine ourselves as having. We have, therefore, to accept whichever proposition we do accept as uninferred. Being uninferred, it appears as, or is, a fact of experience ; and it is as a fact of our experience that we accept it. There is no logical constraint upon us : we are free to accept or reject it. Atheism is possible because the will is free to believe or not believe in God ; and it is because the will is free, and only because it is free, that religion is possible. Hence it is that the common

experience, in which all finite spirits share,
has no constraining power to compel us to
accept either alternative. We are free, we
have freedom to close our eyes to facts, to the
facts of our common experience. If we had
no such freedom, atheism would be unknown,
and the very notion of it an idea that we
simply could not grasp. But we are free : we
are also free to open our eyes and to see the
loving-kindness of the Lord. This freedom is
not a thing which any man is debarred from,
though any man may debar himself from it,
for he is free to deny the freedom of the will
and to assert that he finds no such freedom in
his experience. Nay ! the very man who
asserts his freedom may be very far from
realising it or understanding the dangers to
which it exposes him. To say that we are
free, that we have perfect freedom to believe
in God, sounds so easy. It is only when we
understand that we are free, equally and perfectly free, to disbelieve, that any difficulty
arises. Thus for a long time, in the days of
one's youth, one may go on in happy ignorance
of this latter side of one's freedom, under the
delusion that it is simply impossible to dis-

believe. In the world beyond one's home, one is told or knows that strange things take place —but then they must be strange people who do them—rational persons all believe. But when one goes into the world, one finds that rational persons do not all believe; and that the people who disbelieve are, or at any rate, so long as you did not know, appeared to be, quite human beings. And this disillusionment, this waking up from the delusion that it is impossible for rational beings and estimable men to disbelieve, is sometimes followed by fatal results: it is discovered that disbelief is not an impossibility—and then freedom (which has perhaps coincided with the first escape from home) is conceived to consist wholly in freedom to disbelieve. You are free to disbelieve, there is nothing to stop you—the old-time arguments at any rate are incapable of giving you pause, you have been through all of them, they have no hold on you or power to check you. Thus you may rush to the extreme which is the very opposite to that from which you started: first disbelief was, now belief is, impossible. You go on in ignorance—no longer happy ignorance—of the fact that you still are free. Indeed, the characteristic of those who are in this state is that they find it just inconceivable that any rational creature *can* believe in religion. They do not believe in, or at any rate they have not realised, the freedom of the will. At least they do not believe that in this respect there *is* any freedom for *rational* creatures: there are things which they *must* believe, or rather disbelieve, if they are rational. The trouble is that they cannot help seeing and admitting to themselves that there are other people—to all appearance, indeed beyond all doubt, rational creatures, or rational enough in other matters—who hold themselves under no compulsion to disbelieve. And the man who gets to this point, and has fairness enough to turn the matter over, is very apt to begin to doubt whether there is any such thing as truth, or to assume the thoroughly illogical position of the Agnostic, viz., that there are some things, science for instance, in which truth can be attained; and other things — the only really interesting things—in which it cannot, and with regard to which there very probably is no truth. At

144

Printed and bound by CPI Group (UK) Ltd, Croydon, CR0 4YY

22/10/2024

01777620-0003

this point there is a ray of hope, the possibility
of escape. The man who begins by sitting on
the fence of Agnosticism, will probably come
down : as he sits there and looks down on the
persons on both sides of the fence, and reflects
that they were free to choose and have chosen
freely whichever side they liked, it may occur
to him that after all he cannot be so very
different from other people—that if they are
free to take a side, he probably is equally free.
But this freedom, which before seemed to him
so easy—so easy that he exercised it first one
way and then the other—now wears another
aspect. It requires, or rather perhaps it consists in an exercise of the will. To maintain
the position you have taken up, to say "so I
will, and so I believe," requires a constant
exercise of will-power. To take up your
position is one thing, to maintain it is
another, for your will is still free—otherwise there would be no relapses from conversion.

We are then free, always free, to assume
either that there is or that there is not a God ;
and it is not until you have made the assumption that you can experience its consequences ;

nor until you have experienced its consequences can you know its value. Without
Faith you cannot feel the consequences of
Faith : only by living the life can you do so.
The consequences to be valued must be felt :
no intellectual discussion can take the place of
the feeling. Feeling and emotion are every
whit as essential as reason to the realisation of
truth. If belief is, as it has been defined to
be, " readiness to act," then the glow of belief,
without which there is no action, is something
other and more than that " pure reason " which
Aristotle says "moves nothing." The glow
of belief not only imparts warmth to action, it
gives light to reason. It gives that light by
which we see that the assumption we originally made by faith is no mere assumption
made privately and individually by us, but the
truth which in the words of Hegel (" Philosophy of Religion," i. 3) is "the substance of
actual existing things." " It may have "—this
image of the Absolute, he says—" it may have
a more or less present vitality and certainty
for the religious and devout mind ; or it may
be represented as something longed and
hoped for, far off and in the future. Still,

it always remains a certainty." It is of
the Absolute that Hegel uses these words.
And if in using the word Absolute, "we
depart from Hegel's own usage," and say
that "the Absolute is not God" but a non-
personal principle of unity analogous to that
found in a college or a club, we are stripping
Faith of its emotional quality and thereby
annihilating it. Faith is not purely intellectual
but also emotional, and the emotion is Love.
A principle of unity does not inspire any
particular love, nor can it feel it or pour it
forth. A fellow-being may do both. But
love, even of our neighbour, is not the same
as love of God ; nor is our neighbour's love
the same as God's love. If by the Absolute
we mean simply a principle of unity, there can
be no love in question. But here again the
appeal is to experience : either we do or we
do not feel God's love for us. And in either
case there is little more to say. To feel it
not, is to feel no need of being grateful, to
have nothing—nothing—to be thankful for.
From the brink of that abyss, let us turn away
to the alternative, to God's love for us. If we
feel it, there is an end to any question whether

the Absolute is personal or not : " I know that my Redeemer liveth."

I may now conclude by summing up this course of lectures on the relation of the Evolution of Religion to the Philosophy of Religion. The theory of the Evolution of Religion is concerned essentially with the task of arranging the common experience of mankind, so far as it is religious, under the forms of Space and Time, and in accordance with the law of Cause and Effect. That becomes quite obvious when we reflect that part of the task of students of the Science of Religion consists in endeavouring to determine whether Religion was or was not preceded by a non - religious period in the development of humanity, and—in either case —what were the causes which produced the result. That was the question which I raised in the first lecture : whether the state of things which we find now in the South-Eastern tribes of Central Australia—a state either religious or approximating on religion—or that which we find among the Northern tribes—a state distinctly less religious, or more remote from religion—is the earlier in time and the cause of the other. And in either case, we are further concerned with the task of determining whether the state of things which we find and believe to have been the earlier in one quarter of the globe, in one region of space, is to be assumed to have been also the earlier in a different quarter of the globe, a different region of space. And however we determine the question, we alike assume that the state of things is the effect of preexisting causes.

Thus the theory of the Evolution of Religion cannot work unless we assume the validity of space, time, and the universality of the law of causation. On the other hand it is not necessary, for the purpose of tracing the Evolution of Religion, that we should assume either that the will is free or that God exists : there are distinguished students of the Science of Religion who are convinced that both these latter assumptions are erroneous, and that the assumptions of the validity of space, time, and causation are correct. There are students, however, who differ from them on both points. The points, therefore, must be discussed ; and the discussion is Philosophy. The theory of Evolution, therefore, is one thing, Philosophy is another. The Evolution of Religion is one thing, and the Philosophy of Religion is another. The theory of Evolution generally—as applied to other things than religion — assumes the validity of space, time, and causation ; and therefore the question whether they are valid is a matter to be discussed by Philosophy generally. The theory of the Evolution of Religion in particular insists that the questions as to the freedom of the will and the existence of God must be relegated to Philosophy — to the Philosophy of Religion in particular.

Whether we turn from Evolution or Evolution of Religion to Philosophy or Philosophy of Religion, the first question that confronts us is whether the theory of Evolution—and science generally—is experience or reflection on experience, whether it is concrete or abstract. And I therefore devoted the second lecture to arguing that science is not the facts with which it deals but an abstraction from them, that the Science of Religion is something very different from religion, and that the theory of the Evolution of Religion is not a religious experience.

When once we recognise the difference between an experience and reflection on that experience — when we see that in reflection or by reflection we dissect experience, and that the dissected creature is not the living creature — then it becomes intelligible that space, time, and causation are forms into which we distribute our dissected experience, the order in which we arrange the words of our translation, an order which is peculiar to the translation and does not belong to the original. That was the sum and substance of the third lecture : the theory of Evolution generally assumes the validity of space and time ; Philosophy has to enquire into the validity of the assumption.

Finally, religion assumes the freedom of the will and the existence of God ; and the Philosophy of Religion has to enquire into the validity of those assumptions. Here, therefore, we come once more to close quarters with the question of the relation of the Philosophy of Religion to the Evolution of Religion. Religion, as I say, assumes the freedom of the will and the existence of God. The theory of Evolution, whether applied to religion or to anything else, refuses to make any such assumptions. Some of the most distinguished students of the Evolution of Religion refuse either to make or to believe in those assumptions. The question as to the validity of these assumptions, which by some are regarded as true and by others as not true, may be discussed; and if it is discussed, the discussion is Philosophy—Philosophy of Religion. The position which I have taken up in this, the fourth lecture, is that, though the question may and must constantly be discussed, it cannot be settled by Philosophy. The business or part of the business of Philosophy— and it has hitherto discharged that part of its business very thoroughly—is to examine attempts to prove that there is or is not a God, and to show that without Faith there is no proof. The whole question is as to the contents of experience; and whether we will see and attend to what is there, is a matter of will — free - will — or perhaps I should rather say that when we have Faith and when we do believe, then we can go on to erect a philosophy on that foundation. Then we are in a position to assert what the content of experience is and to build upon it. It is upon experience that constructive philosophy must be based. If the work of excavating and clearing out the foundations appears to be destructive, let us remember that it is destructive only if we propose to erect nothing on the site ; and that the appearance is wholly misleading if our hope is there to raise a temple of the Lord.

The whole question is, I say, as to the content of experience — whether we have experience of God. If we are satisfied that we have, then our position is that those who believe they have not, believe so freely, of their own free will ; and that they are mistaken in so believing. They base themselves on their own personal experience ; and, we cannot help believing, they tend to overlook the fact that in so doing they fall into the fallacy of Solipsism, the fallacy, that is, of imagining that experience is limited to myself, the fallacy of denying that there is a common experience in which all finite spirits share, and of which no individual is the sole, authorised interpreter. On the other hand, there are those who endeavour to make this community or communion of spirits their starting-point, and to argue from it to the Absolute, the principle of unity. By some of them it is maintained that this principle of unity can be proved to be a personal God; by others it is maintained that it can only be an impersonal principle, and that therefore there is no God. On the whole, therefore, I think it wiser to regard these arguments as an additional proof of the position that we must begin with belief in God, by recognising that He is no inference or hypothesis, and that we must not hope to reach Him by putting a train of reasoning between us and Him. At that distance spiritual communion becomes a remote possibility, whereas we know it in experience as the most immediate fact.